Multivariate Statistical Methods
A Primer
Fourth Edition

Multivariate Statistical Methods
A Primer

Fourth Edition

Bryan F. J. Manly

University of Otago
Dunedin, New Zealand

Jorge A. Navarro Alberto

Universidad Autónoma de Yucatán
Mérida, México

CRC Press
Taylor & Francis Group
Boca Raton London New York

CRC Press is an imprint of the
Taylor & Francis Group, an **informa** business

A CHAPMAN & HALL BOOK

CRC Press
Taylor & Francis Group
6000 Broken Sound Parkway NW, Suite 300
Boca Raton, FL 33487-2742

International Standard Book Number-13: 978-1-4987-2896-6 (Paperback)

Library of Congress Cataloging-in-Publication Data

Names: Manly, Bryan F. J., 1944- | Navarro Alberto, Jorge A., 1963-
Title: Multivariate statistical methods : a primer.
Description: Fourth edition / Bryan F.J. Manly and Jorge A. Navarro Alberto.
| Boca Raton : CRC Press, 2017. | Includes bibliographical references and index.
Identifiers: LCCN 2016020181 | ISBN 9781498728966
Subjects: LCSH: Multivariate analysis.
Classification: LCC QA278 .M35 2017 | DDC 519.5/35--dc23
LC record available at https://lccn.loc.gov/2016020181

Visit the Taylor & Francis Web site at
http://www.taylorandfrancis.com

and the CRC Press Web site at
http://www.crcpress.com

Dedication

*To my multivariate (8-dimensional) space of caring
and loving support: the Navarro-Contreras*

J.N.A.

A journey of a thousand miles begins with a single step

Lao Tsu

Contents

Preface

This is the fourth edition of the book *Multivariate Statistical Methods: a Primer*. The contents are similar to what was in the third edition of the book, with the main difference being the introduction of R code to do all of the analyses in the fourth edition. The version of R used for running the R-scripts (and the corresponding packages) is R 3.3.1. Also, the results obtained with the R code have been checked to ensure that they are the same as the results obtained from various other statistical packages.

The purpose of the book is to introduce multivariate statistical methods to nonmathematicians. It is not intended to be comprehensive. Rather, the intention is to keep the details to a minimum while still conveying a good idea about what can be done. In other words, it is a book to "get you going" in a particular area of statistical methods.

It is assumed that readers have a working knowledge of elementary statistics, including tests of significance using the normal, t, chi-squared, and F distributions, analysis of variance, and linear regression. The material covered in a typical first-year university course in statistics should be quite adequate in this respect. Some facility with algebra is also required to follow the equations in certain parts of the text, and understanding the theory of multivariate methods to some extent does require the use of matrix algebra. However, the amount needed is not great if some details are accepted on faith. Matrix algebra is summarized in Chapter 2, and anyone who masters this chapter will have a reasonable competency in this area.

To some extent, the chapters can be read independently of each other. The first five are preliminary reading, because they are mainly concerned with general aspects of multivariate data rather than with specific techniques. Chapter 1 introduces some examples with the aim of motivating the analyses covered in the book. Chapter 2 covers matrix algebra, Chapter 3 is about graphical methods of various types, Chapter 4 is about tests of significance, and Chapter 5 is about the measurement of distances between objects based on variables measured on those objects. It is recommended that these chapters should be reviewed before Chapters 6 through 12, which cover the most important multivariate procedures in current use. The final Chapter 13 contains some general comments about the analysis of multivariate data.

The chapters in this fourth edition of the book are the same as those in the second and third editions, and in making changes we have continually kept in mind the original intention of the book, which was that it should be as short as possible and attempt to do no more than take readers to the stage where they can begin to use multivariate methods in an intelligent manner. An Appendix to Chapter 1 provides an introduction to the use of the R package, and the code for analyses is discussed in Appendices for Chapters 2 through 12.

We wish to thank the staff of Chapman and Hall/CRC Press for their work over the years in promoting this book and encouraging us to produce this fourth edition.

Bryan F.J. Manly
University of Otago
Dunedin, New Zealand

Jorge A. Navarro Alberto
Universidad Autónoma de Yucatán
Mérida, Mexico

Authors

Bryan F.J. Manly was a professor of statistics at the University of Otago, Dunedin, New Zealand until 2000 after which he moved to the United States to work as a consultant for Western EcoSystems Technology Inc. In 2015 he returned to New Zealand and is now again a professor at the University of Otago in the School of Medicine.

Jorge A. Navarro Alberto is a mathematician and professor at the Autonomous University of Yucatán, Mexico, where he has taught statistics for biologists, marine biologists and natural resource management students for more than 25 years. His research interests are centered in ecological and environmental statistics, primarily in statistical modeling of biological data, multivariate statistical methods applied to the study of ecological communities, ecological sampling, and null models in ecology.

chapter one

The material of
multivariate analysis

1.1 Examples of multivariate data

The statistical methods that are described in elementary texts are mostly
univariate methods, because they are only concerned with analyzing
variation in a single random variable. On the other hand, the whole point
of a multivariate analysis is to consider several related variables simultaneously, with each one being considered to be equally important, at least
initially. The potential value of this more general approach can be seen by
considering a few examples.

Example 1.1: Storm survival of sparrows

After a severe storm on February 1, 1898, a number of moribund sparrows were taken to Hermon Bumpus' biological laboratory at Brown
University, Rhode Island. Subsequently, about half of the birds died,
and Bumpus saw this as an opportunity to see whether he could find
any support for Charles Darwin's theory of natural selection. To this
end, he made eight morphological measurements on each bird and
also weighed the birds. The results for five of the measurements are
shown in Table 1.1, for females only.

From the data that he obtained, Bumpus (1898) concluded
"[that] the birds which perished, perished not through accident,
but because they were physically disqualified, and that the birds
which survived, survived because they possessed certain physical
characters." Specifically, he found that the survivors "are shorter
and weigh less ... have longer wing bones, longer legs, longer
sternums and greater brain capacity" than the nonsurvivors. He
also concluded that "the process of selective elimination is most
severe with extremely variable individuals, no matter in which
direction the variation may occur. It is quite as dangerous to be
above a certain standard of organic excellence as it is to be con-
spicuously below the standard." This was to say that stabilizing
selection occurred, so that individuals with measurements close to
the average survived better than individuals with measurements
far from the average.

In fact, the development of multivariate statistical methods had
hardly begun in 1898, when Bumpus was writing. The correlation

Table 1.1 Body measurements of female sparrows

Bird	X_1	X_2	X_3	X_4	X_5
1	156	245	31.6	18.5	20.5
2	154	240	30.4	17.9	19.6
3	153	240	31.0	18.4	20.6
4	153	236	30.9	17.7	20.2
5	155	243	31.5	18.6	20.3
6	163	247	32.0	19.0	20.9
7	157	238	30.9	18.4	20.2
8	155	239	32.8	18.6	21.2
9	164	248	32.7	19.1	21.1
10	158	238	31.0	18.8	22.0
11	158	240	31.3	18.6	22.0
12	160	244	31.1	18.6	20.5
13	161	246	32.3	19.3	21.8
14	157	245	32.0	19.1	20.0
15	157	235	31.5	18.1	19.8
16	156	237	30.9	18.0	20.3
17	158	244	31.4	18.5	21.6
18	153	238	30.5	18.2	20.9
19	155	236	30.3	18.5	20.1
20	163	246	32.5	18.6	21.9
21	159	236	31.5	18.0	21.5
22	155	240	31.4	18.0	20.7
23	156	240	31.5	18.2	20.6
24	160	242	32.6	18.8	21.7
25	152	232	30.3	17.2	19.8
26	160	250	31.7	18.8	22.5
27	155	237	31.0	18.5	20.0
28	157	245	32.2	19.5	21.4
29	165	245	33.1	19.8	22.7
30	153	231	30.1	17.3	19.8
31	162	239	30.3	18.0	23.1
32	162	243	31.6	18.8	21.3
33	159	245	31.8	18.5	21.7
34	159	247	30.9	18.1	19.0
35	155	243	30.9	18.5	21.3
36	162	252	31.9	19.1	22.2
37	152	230	30.4	17.3	18.6
38	159	242	30.8	18.2	20.5
39	155	238	31.2	17.9	19.3

Table 1.1 (Continued) Body measurements of female sparrows

Bird	X_1	X_2	X_3	X_4	X_5
40	163	249	33.4	19.5	22.8
41	163	242	31.0	18.1	20.7
42	156	237	31.7	18.2	20.3
43	159	238	31.5	18.4	20.3
44	161	245	32.1	19.1	20.8
45	155	235	30.7	17.7	19.6
46	162	247	31.9	19.1	20.4
47	153	237	30.6	18.6	20.4
48	162	245	32.5	18.5	21.1
49	164	248	32.3	18.8	20.9

Source: Data from Bumpus, H.C., *Biological Lectures*, Marine Biology Laboratory, Woods Hole, MA, 1898.

Note: X_1 = total length, X_2 = alar extent, X_3 = length of beak and head, X_4 = length of humerus, and X_5 = length of keel of sternum (all in millimeters). Original measurements were in inches and millimeters. Birds 1–21 survived and birds 22–49 died.

coefficient as a measure of the relationship between two variables was devised by Francis Galton in 1877. However, it was another 56 years before Harold Hotelling described a practical method for carrying out a principal components analysis, which is one of the simplest multivariate analyses that can usefully be applied to Bumpus' data. Bumpus did not even calculate standard deviations. Nevertheless, his methods of analysis were sensible. Many authors have reanalyzed his data and, in general, have confirmed his conclusions.

Taking the data as an example for illustrating multivariate methods, several interesting questions arise, in particular:

1. How are the various measurements related? For example, does a large value for one of the variables tend to occur with large values for the other variables?
2. Are there statistically significant differences between survivors and nonsurvivors for the mean values of the variables?
3. Do the survivors and nonsurvivors show similar amounts of variation for the variables?
4. If the survivors and nonsurvivors do differ in terms of the distributions of the variables, then is it possible to construct some function of these variables that separates the two groups? It would then be convenient if large values of the function tended to occur with the survivors, as the function would then apparently be an index of the Darwinian fitness of the sparrows.

Example 1.2: Egyptian skulls

For a second example, consider the data shown in Table 1.2 for measurements made on male skulls from the area of Thebes in Egypt. There are five samples of 30 skulls from each of the early predynastic period (circa 4000 BC), the late predynastic period (circa 3300 BC), the 12th and 13th Dynasties (circa 1850 BC), the Ptolemaic period (circa 200 BC), and the Roman period (circa AD 150). Four measurements are available on each skull, as illustrated in Figure 1.1.

For this example, some interesting questions are

1. How are the four measurements related?
2. Are there statistically significant differences between the sample means for the variables, and if so, do these differences reflect gradual changes over time in the shape and size of skulls?
3. Are there significant differences between the sample standard deviations for the variables, and if so, do these differences reflect gradual changes over time in the amount of variation?
4. Is it possible to construct a function of the four variables that in some sense describes the changes over time?

These questions are, of course, rather similar to the ones suggested for Example 1.1.

It will be seen later that there are differences between the five samples that can be explained partly as time trends. It must be said, however, that the reasons for the apparent changes are unknown. Migration of other races into the region may well have been the most important factor.

Example 1.3: Distribution of a butterfly

A study of 16 colonies of the butterfly *Euphydryas editha* in California and Oregon produced the data shown in Table 1.3. Here, there are four environmental variables (altitude, annual precipitation, and the minimum and maximum temperatures) and six genetic variables (percentage frequencies for different Pgi genes as determined by the technique of electrophoresis). For the purposes of this example, there is no need to go into the details of how the gene frequencies were determined, and strictly speaking, they are not exactly gene frequencies anyway. It is sufficient to say that the frequencies describe the genetic distribution of the butterfly to some extent. Figure 1.2 shows the geographical locations of the colonies.

In this example, questions that can be asked include

1. Are the Pgi frequencies similar for colonies that are close in space?
2. To what extent, if any, are the Pgi frequencies related to the environmental variables?

Table 1.2 Measurement on male Egyptian skulls

Skull	Early predynastic				Late predynastic				12th and 13th Dynasties				Ptolemaic period				Roman period			
	X_1	X_2	X_3	X_4	X_1	X_2	X_3	X_4	X_1	X_2	X_3	X_4	X_1	X_2	X_3	X_4	X_1	X_2	X_3	X_4
1	131	138	89	49	124	138	101	48	137	141	96	52	137	134	107	54	137	123	91	50
2	125	131	92	48	133	134	97	48	129	133	93	47	141	128	95	53	136	131	95	49
3	131	132	99	50	138	134	98	45	132	138	87	48	141	130	87	49	128	126	91	57
4	119	132	96	44	148	129	104	51	130	134	106	50	135	131	99	51	130	134	92	52
5	136	143	100	54	126	124	95	45	134	134	96	45	133	120	91	46	138	127	86	47
6	138	137	89	56	135	136	98	52	140	133	98	50	131	135	90	50	126	138	101	52
7	139	130	108	48	132	145	100	54	138	138	95	47	140	137	94	60	136	138	97	58
8	125	136	93	48	133	130	102	48	136	145	99	55	139	130	90	48	126	126	92	45
9	131	134	102	51	131	134	96	50	136	131	92	46	140	134	90	51	132	132	99	55
10	134	134	99	51	133	125	94	46	126	136	95	56	138	140	100	52	139	135	92	54
11	129	138	95	50	133	136	103	53	137	129	100	53	132	133	90	53	143	120	95	51
12	134	121	95	53	131	139	98	51	137	139	97	50	134	134	97	54	141	136	101	54
13	126	129	109	51	131	136	99	56	136	126	101	50	135	135	99	50	135	135	95	56
14	132	136	100	50	138	134	98	49	137	133	90	49	133	136	95	52	137	134	93	53
15	141	140	100	51	130	136	104	53	129	142	104	47	136	130	99	55	142	135	96	52
16	131	134	97	54	131	128	98	45	135	138	102	55	134	137	93	52	139	134	95	47
17	135	137	103	50	138	129	107	53	129	135	92	50	131	141	99	55	138	125	99	51
18	132	133	93	53	123	131	101	51	134	125	90	60	129	135	95	47	137	135	96	54
19	139	136	96	50	130	129	105	47	138	134	96	51	136	128	93	54	133	125	92	50
20	132	131	101	49	134	130	93	54	136	135	94	53	131	125	88	48	145	129	89	47

(*Continued*)

Table 1.2 (Continued) Measurement on male Egyptian skulls

Skull	Early predynastic				Late predynastic				12th and 13th Dynasties				Ptolemaic period				Roman period			
	X_1	X_2	X_3	X_4	X_1	X_2	X_3	X_4	X_1	X_2	X_3	X_4	X_1	X_2	X_3	X_4	X_1	X_2	X_3	X_4
21	126	133	102	51	137	136	106	49	132	130	91	52	139	130	94	53	138	136	92	46
22	135	135	103	47	126	131	100	48	133	131	100	50	144	124	86	50	131	129	97	44
23	134	124	93	53	135	136	97	52	138	137	94	51	141	131	97	53	143	126	88	54
24	128	134	103	50	129	126	91	50	130	127	99	45	130	131	98	53	134	124	91	55
25	130	130	104	49	134	139	101	49	136	133	91	49	133	128	92	51	132	127	97	52
26	138	135	100	55	131	134	90	53	134	123	95	52	138	126	97	54	137	125	85	57
27	128	132	93	53	132	130	104	50	136	137	101	54	131	142	95	53	129	128	81	52
28	127	129	106	48	130	132	93	52	133	131	96	49	136	138	94	55	140	135	103	48
29	131	136	114	54	135	132	98	54	138	133	100	55	132	136	92	52	147	129	87	48
30	124	138	101	46	130	128	101	51	138	133	91	46	135	130	100	51	136	133	97	51

Source: Data from Thomson, A. and Randall-Maciver, P., *Ancient Races of the Thebaid*, Oxford University Press, Oxford, London, 1905.

X_1 = maximum breadth, X_2 = basibregmatic height, X_3 = basialveolar length, X_4 = nasal height, all in millimeters.

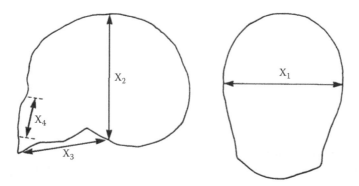

Figure 1.1 Four measurements made on Egyptian male skulls.

These are important questions in trying to decide how the Pgi frequencies are determined. If the genetic composition of the colonies was largely determined by past and present migration, then gene frequencies will tend to be similar for colonies that are close in space, but may show little relationship with the environmental variables. On the other hand, if it is the environment that is most important, then this should show up in relationships between the gene frequencies and the environmental variables (assuming that the right variables have been measured), but close colonies will only have similar gene frequencies if they have similar environments. Obviously, colonies that are close in space will usually have similar environments, so that it may be difficult to reach a clear conclusion on this matter.

Example 1.4: Prehistoric dogs from Thailand

Excavations of prehistoric sites in northeast Thailand have produced a collection of canid (dog) bones covering a period from about 3500 BC to the present. However, the origin of the prehistoric dog is not certain. It may have descended from the golden jackal (*Canis aureus*) or from the wolf, but the wolf is not native to Thailand. The nearest indigenous sources are western China (*Canis lupus chanco*) and the Indian subcontinent (*Canis lupus pallides*).

To try to clarify the ancestors of the prehistoric dogs, mandible (lower jaw) measurements were made on the available specimens. These were then compared with the same measurements made on the golden jackal, the Chinese wolf, and the Indian wolf. The comparisons were also extended to include the dingo, which may have its origins in India; the cuon (*Cuon alpinus*), which is indigenous to Southeast Asia, and modern village dogs from Thailand.

Table 1.4 gives mean values for the six mandible measurements for specimens from all seven groups. The main question here is what the measurements suggest about the relationships between the groups and, in particular, how the prehistoric dog seems to relate to the other groups.

Table 1.3 Environmental variables and phosphoglucose-isomerase (Pgi) gene frequencies for colonies of the butterfly *Euphydryas editha* in California and Oregon, USA

Colony	Altitude (feet)	Annual precipitation (inches)	Temperature (°F)		Frequencies of Pgi mobility genes (%)[a]					
			Maximum	Minimum	0.4	0.6	0.8	1	1.16	1.3
SS	500	43	98	17	0	3	22	57	17	1
SB	808	20	92	32	0	16	20	38	13	13
WSB	570	28	98	26	0	6	28	46	17	3
JRC	550	28	98	26	0	4	19	47	27	3
JRH	550	28	98	26	0	1	8	50	35	6
SJ	380	15	99	28	0	2	19	44	32	3
CR	930	21	99	28	0	0	15	50	27	8
UO	650	10	101	27	10	21	40	25	4	0
LO	600	10	101	27	14	26	32	28	0	0
DP	1,500	19	99	23	0	1	6	80	12	1
PZ	1,750	22	101	27	1	4	34	33	22	6
MC	2,000	58	100	18	0	7	14	66	13	0
IF	2,500	34	102	16	0	9	15	47	21	8
AF	2,000	21	105	20	3	7	17	32	27	14
GH	7,850	42	84	5	0	5	7	84	4	0
GL	10,500	50	81	−12	0	3	1	92	4	0

Source: Data from McKechnie, S.W. et al., *Genetics*, 81, 571–94, 1975, with the environmental variables rounded to integers for simplicity.

Note: The original data were for 21 colonies, but for the present example, five colonies with small samples for the estimation of gene frequencies have been excluded to make all estimates about equally reliable.

[a] The numbers 0.40, 0.60, etc. represent different genetic types of Pgi, so that the frequencies for a colony (adding to 100%) show the frequencies of the different types for the *E. editha* at that location.

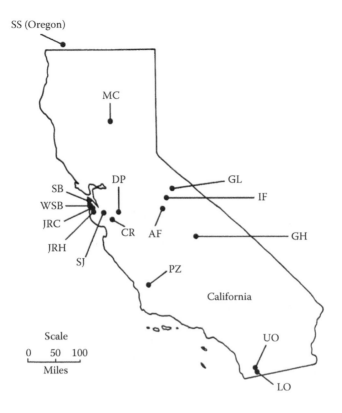

SS (Oregon)

MC

DP

GL

SB

IF

WSB

JRC

CR   AF

GH

JRH

SJ

PZ

California

Scale

UO

0   50   100

Miles

LO

Figure 1.2 Colonies of *Euphydras editha* in California and Oregon.

Table 1.4 Mean mandible measurements for seven canine groups

Group	X_1	X_2	X_3	X_4	X_5	X_6
Modern dog	9.7	21.0	19.4	7.7	32.0	36.5
Golden jackal	8.1	16.7	18.3	7.0	30.3	32.9
Chinese wolf	13.5	27.3	26.8	10.6	41.9	48.1
Indian wolf	11.5	24.3	24.5	9.3	40.0	44.6
Cuon	10.7	23.5	21.4	8.5	28.8	37.6
Dingo	9.6	22.6	21.1	8.3	34.4	43.1
Prehistoric dog	10.3	22.1	19.1	8.1	32.2	35

Source: Data from Higham, C.F.W. et al., *J. Archaeol. Sci.*, 7, 149–65, 1980.

Note: X_1 = breadth of mandible, X_2 = height of mandible below the first molar, X_3 = length of the first molar, X_4 = breadth of the first molar, X_5 = length from first to third molar inclusive, and X_6 = length from first to fourth premolar inclusive (all in millimeters).

Example 1.5: Employment in European countries

Finally, as a contrast to the previous biological examples, consider the data in Table 1.5. This shows the percentages of the labor force in nine different types of industry for 30 European countries. In this case, multivariate methods may be useful in isolating groups of countries with similar employment patterns, and in generally aiding the understanding of the relationships between the countries. Differences between countries that are related to political grouping (the European Union [EU]; the European Free Trade Area [EFTA]; the Eastern European countries; and other countries) may be of particular interest.

1.2 Preview of multivariate methods

The five examples just considered are typical of the raw material for multivariate statistical methods. In all cases, there are several variables of interest, and these are clearly not independent of each other. At this point, it is useful to give a brief preview of what is to come in the chapters that follow in relationship to these examples.

Principal components analysis is designed to reduce the number of variables that need to be considered to a small number of indices (called the principal components) that are linear combinations of the original variables. For example, much of the variation in the body measurements of sparrows (X_1–X_5) shown in Table 1.1 will be related to the general size of the birds, and the total

$$I_1 = X_1 + X_2 + X_3 + X_4 + X_5$$

should measure this aspect of the data quite well. This accounts for one dimension of the data. Another index is

$$I_2 = X_1 + X_2 + X_3 - X_4 - X_5,$$

which is a contrast between the first three measurements and the last two. This reflects another dimension of the data. Principal components analysis provides an objective way of finding indices of this type so that the variation in the data can be accounted for as concisely as possible. It may well turn out that two or three principal components provide a good summary of all the original variables. Consideration of the values of the principal components instead of the values of the original variables may then make it much easier to understand what the data have to say. In short, principal components analysis is a means of simplifying data by reducing the number of variables.

Table 1.5 Percentages of the workforce employed in nine different industry
groups in 30 countries in Europe

Country	Group	AGR	MIN	MAN	PS	CON	SER	FIN	SPS	TC
Belgium	EU	2.6	0.2	20.8	0.8	6.3	16.9	8.7	36.9	6.8
Denmark	EU	5.6	0.1	20.4	0.7	6.4	14.5	9.1	36.3	7.0
France	EU	5.1	0.3	20.2	0.9	7.1	16.7	10.2	33.1	6.4
Germany	EU	3.2	0.7	24.8	1.0	9.4	17.2	9.6	28.4	5.6
Greece	EU	22.2	0.5	19.2	1.0	6.8	18.2	5.3	19.8	6.9
Ireland	EU	13.8	0.6	19.8	1.2	7.1	17.8	8.4	25.5	5.8
Italy	EU	8.4	1.1	21.9	0.0	9.1	21.6	4.6	28.0	5.3
Luxembourg	EU	3.3	0.1	19.6	0.7	9.9	21.2	8.7	29.6	6.8
Netherlands	EU	4.2	0.1	19.2	0.7	0.6	18.5	11.5	38.3	6.8
Portugal	EU	11.5	0.5	23.6	0.7	8.2	19.8	6.3	24.6	4.8
Spain	EU	9.9	0.5	21.1	0.6	9.5	20.1	5.9	26.7	5.8
United Kingdom	EU	2.2	0.7	21.3	1.2	7.0	20.2	12.4	28.4	6.5
Austria	EFTA	7.4	0.3	26.9	1.2	8.5	19.1	6.7	23.3	6.4
Finland	EFTA	8.5	0.2	19.3	1.2	6.8	14.6	8.6	33.2	7.5
Iceland	EFTA	10.5	0.0	18.7	0.9	10.0	14.5	8.0	30.7	6.7
Norway	EFTA	5.8	1.1	14.6	1.1	6.5	17.6	7.6	37.5	8.1
Sweden	EFTA	3.2	0.3	19.0	0.8	6.4	14.2	9.4	39.5	7.2
Switzerland	EFTA	5.6	0.0	24.7	0.0	9.2	20.5	10.7	23.1	6.2
Albania	Eastern	55.5	19.4	0.0	0.0	3.4	3.3	15.3	0.0	3.0
Bulgaria	Eastern	19.0	0.0	35.0	0.0	6.7	9.4	1.5	20.9	7.5
Czech/Slovak Reps	Eastern	12.8	37.3	0.0	0.0	8.4	10.2	1.6	22.9	6.9
Hungary	Eastern	15.3	28.9	0.0	0.0	6.4	13.3	0.0	27.3	8.8
Poland	Eastern	23.6	3.9	24.1	0.9	6.3	10.3	1.3	24.5	5.2
Romania	Eastern	22.0	2.6	37.9	2.0	5.8	6.9	0.6	15.3	6.8
USSR (Former)	Eastern	18.5	0.0	28.8	0.0	10.2	7.9	0.6	25.6	8.4
Yugoslavia (Former)	Eastern	5.0	2.2	38.7	2.2	8.1	13.8	3.1	19.1	7.8
Cyprus	Other	13.5	0.3	19.0	0.5	9.1	23.7	6.7	21.2	6.0
Gibraltar	Other	0.0	0.0	6.8	2.0	16.9	24.5	10.8	34.0	5.0
Malta	Other	2.6	0.6	27.9	1.5	4.6	10.2	3.9	41.6	7.2
Turkey	Other	44.8	0.9	15.3	0.2	5.2	12.4	2.4	14.5	4.4

Source: Data from Euromonitor (1995), except for Germany and the United Kingdom,
where more recent values were obtained from the United Nations *Statistical
Yearbook*, 2000.

Note: AGR, agriculture, forestry, and fishing; MIN, mining and quarrying; MAN, manufacturing; PS, power and water supplies; CON, construction; SER, services; FIN, finance;
SPS, social and personal services; TC, transport and communications.

Factor analysis also attempts to account for the variation in a number of original variables using a smaller number of index variables or factors. It is assumed that each original variable can be expressed as a linear combination of these factors, plus a residual term that reflects the extent to which the variable is independent of the other variables. For example, a two-factor model for the sparrow data assumes that

$$X_1 = a_{11}F_1 + a_{12}F_2 + e_1$$
$$X_2 = a_{21}F_1 + a_{22}F_2 + e_2$$

$$X_3 = a_{31}F_1 + a_{32}F_2 + e_3$$
$$X_4 = a_{41}F_1 + a_{42}F_2 + e_4$$

and

$$X_5 = a_{51}F_1 + a_{52}F_2 + e_5$$

where:
   a_{ij} values are constants
   F_1 and F_2 are the factors
   e_i represents the variation in X_i that is independent of the variation in
       the other X variables

Here, F_1 might be the factor of size. In that case, the coefficients a_{11}, a_{21}, a_{31}, a_{41}, and a_{51} would all be positive, reflecting the fact that some birds tend to be large and some birds tend to be small on all body measurements. The second factor F_2 might then measure an aspect of the shape of birds, with some positive coefficients and some negative coefficients. If this two-factor model fitted the data well, then it would provide a relatively straightforward description of the relationship between the five body measurements being considered.

One type of factor analysis starts by taking the first few principal components as the factors in the data being considered. These initial factors are then modified by a special transformation process called *factor rotation* to make them easier to interpret. Other methods for finding initial factors are also used. A rotation to simpler factors is almost always done.

Discriminant function analysis is concerned with the problem of seeing whether it is possible to separate different groups on the basis of the available measurements. This could be used, for example, to see how well surviving and nonsurviving sparrows can be separated using their body measurements (Example 1.1) or how skulls from different epochs can be separated, again using size measurements (Example 1.2). Like principal components analysis, discriminant function analysis is based on the idea of finding suitable linear combinations of the original variables to achieve the intended aim.

Cluster analysis is concerned with the identification of groups of similar objects. There is not much point in doing this type of analysis with data like those of Examples 1.1 and 1.2, as the groups (survivors/nonsurvivors and epochs) are already known. However, in Example 1.3, there might be some interest in grouping colonies on the basis of environmental variables or Pgi frequencies, while in Example 1.4, the main point of interest is in the similarity between prehistoric Thai dogs and other animals. Likewise, in Example 1.5, the European countries can possibly be grouped in terms of their similarity in employment patterns.

With *canonical correlation*, the variables (not the objects) are divided into two groups, and interest centers on the relationship between these. Thus, in Example 1.3, the first four variables are related to the environment, while the remaining six variables reflect the genetic distribution at the different colonies of *Euphydryas editha*. Finding what relationships, if any, exist between these two groups of variables is of considerable biological interest.

Multidimensional scaling begins with data on some measure of the distances between a number of objects. From these distances, a map is then constructed showing how the objects are related. This is a useful facility, as it is often possible to measure how far apart pairs of objects are without having any idea of how the objects are related in a geometric sense. Thus, in Example 1.4, there are ways of measuring the distances between modern dogs and golden jackals, modern dogs and Chinese wolves, and so on. Considering each pair of animal groups gives 21 distances altogether, and from these distances, multidimensional scaling can be used to produce a type of map of the relationships between the groups. With a one-dimensional map, the groups are placed along a straight line. With a two-dimensional map, they are represented by points on a plane. With a three-dimensional map, they are represented by points within a cube. Four-dimensional and higher solutions are also possible, although these have limited use because they cannot be visualized in a simple way. The value of a one-, two-, or three-dimensional map is clear for Example 1.4, as such a map would immediately show the groups to which prehistoric dogs are most similar. Hence, multidimensional scaling may be a useful alternative to cluster analysis in this case. A map of European countries based on their employment patterns might also be of interest in Example 1.5.

Principal components analysis and multidimensional scaling are sometimes referred to as methods for *ordination*. That is to say, they are methods for producing axes against which a set of objects of the interest can be plotted. Other methods of ordination are also available.

Principal coordinates analysis is like a type of principal components analysis that starts off with information on the extent to which the pairs of objects are different in a set of objects, instead of the values for measurements on the objects. As such, it is intended to do the same as multidimensional scaling. However, the assumptions made and the numerical methods used are not the same.

Correspondence analysis starts with data on the abundance of each of several characteristics for each of a set of objects. This is useful in ecology, for example, where the objects of interest are often different sites, the characteristics are different species, and the data consist of abundances of the species in samples taken from the sites. The purpose of correspondence analysis would then be to clarify the relationships between the sites, as expressed by species distributions, and the relationships between the species, as expressed by site distributions.

1.3   The multivariate normal distribution

The normal distribution for a single variable should be familiar to readers of this book. It has the well-known bell-shaped frequency curve, and many standard univariate statistical methods are based on the assumption that data are normally distributed.

Knowing the prominence of the normal distribution with univariate statistical methods, it will come as no surprise to discover that the multivariate normal distribution has a central position with multivariate statistical methods. Many of these methods require the assumption that the data being analyzed have multivariate normal distributions.

The exact definition of a multivariate normal distribution is not too important. The approach of most people, for better or worse, seems to be to regard data as being normally distributed unless there is some reason to believe that this is not true. In particular, if all the individual variables being studied appear to be normally distributed, then it is assumed that the joint distribution is multivariate normal. This is, in fact, a minimum requirement, because the definition of multivariate normality requires more than this.

Cases do arise where the assumption of multivariate normality is clearly invalid. For example, one or more of the variables being studied may have a highly skewed distribution with several very high (or low) values, there may be many repeated values, and so on. This type of problem can sometimes be overcome by an appropriate transformation of the data, as discussed in elementary texts on statistics. If this does not work, then a rather special form of analysis may be required.

One important aspect of a multivariate normal distribution is that it is specified completely by a mean vector and a covariance matrix. The definitions of a mean vector and a covariance matrix are given in Section 2.7. Basically, the mean vector contains the mean values for all the variables being considered, while the covariance matrix contains the variances for all the variables plus the covariances, which measure the extent to which all pairs of variables are related.

1.4 Computer programs

Practical methods for carrying out the calculations for multivariate analyses have been developed over about the last 80 years. However, the application of these methods for more than a small number of variables had to wait until computers became readily available. Therefore, it is only in the last 40 years or so that the methods have become reasonably easy to carry out for the average researcher.

Nowadays, there are many standard statistical packages and computer programs available for calculations on computers of all types. It is intended that this book should provide readers with enough information to use any of these packages and programs intelligently. However, given the wide accessibility of the R programming language, we have emphasized its use. The Appendix to this chapter therefore gives a brief review of the R environment needed to run the R commands included at the end of the following chapters. The R codes for most of the examples are also available at the website http://www.manly-biostatistics.co.nz/. More extensive treatments of the R language can be found in many books and manuals, including those by Adler (2012), Crawley (2013), Logan (2010), Teetor (2011), and Venables et al. (2015). Additional references on the use of R for multivariate analysis are also provided in the chapters that follow.

References

Adler, J. (2012). *R in a Nutshell*. 2nd Edn. Sebastopol, CA: O'Reilly Media.

Bumpus, H.C. (1898). The elimination of the unfit as illustrated by the introduced sparrow, *Passer domesticus. Biological Lectures*, 11th Lecture. Marine Biology Laboratory, Woods Hole, MA, 209–26.

Crawley, M. (2013). *The R Book*. 2nd Edn. Chichester: Wiley.

Euromonitor (1995). *European Marketing Data and Statistics*. London: Euromonitor Publications.

Higham, C.F.W., Kijngam, A., and Manly, B.F.J. (1980). An analysis of prehistoric canid remains from Thailand. *Journal of Archaeological Science* 7: 149–65.

Logan, M. (2010). *Biostatistical Design and Analysis Using R: A Practical Guide*. Chichester: Wiley.

McKechnie, S.W., Ehrlich, P.R., and White, R.R. (1975). Population genetics of Euphydryas butterflies. I. Genetic variation and the neutrality hypothesis. *Genetics* 81: 571–94.

Teetor, P. (2011). *R Cookbook*. Sebastopol, CA: O'Reilly Media.

Thomson, A. and Randall-Maciver, R. (1905). *Ancient Races of the Thebaid*. Oxford, London: Oxford University Press.

United Nations (2000). *Statistical Yearbook*, 44th Issue. New York: Department of Social Affairs.

Venables, W.N., Smith, D.M. and the R Core Team. (2015). *An Introduction to R: Notes on R, A Programming Environment for Data Analysis and Graphics, Version 3.1.3 (2015-03-09)*. http://cran.r-project.org/doc/manuals/R-intro.pdf

Appendix: An Introduction to R

In the website of the Comprehensive R Archive Network (CRAN, http://www.r-project.org), R is described as "a *free* software environment for statistical computing and graphics ... maintained by a R Development Core Team." Nowadays, R has taken a leading role in science, business, and technology as a computational system for excellence with data manipulation, calculation of statistical procedures, and graphical displays. You can install R in any operating system (Windows, Mac, or Linux). If you have a Mac or a Windows system, you may want to execute the installer, downloaded from the mirror site of your preference. You just need to follow the directions given in www.r-project.org. Usually, a new version of R is published in March and September each year, but you should be aware of the latest releases announced in the News section of the r-project website.

A.1 The R graphical user interface

R has a simple graphical user interface (GUI) that is adequate to input code, to get numerical results displayed in a console, and to generate new pop-up windows (the *Graphics devices*) where graphics are produced. Once R is started, you will see the R Console window containing information about the version of R you are running and other aspects of R. Then, the prompt > appears, indicating that R is in interactive mode. This means that R is waiting for a command to be written and executed. Another way to work in R, the most usual alternative, is through the script editor, accessible from the File menu. Whenever the script editor is invoked, a simple text editor pops up, so that you are able to write sequences of R commands, with the advantage that they can be saved in an ASCII file, readable by any text editor. By using the procedures Run line or selection or Run all, found in the Edit menu, sets of R code or the entire script can be passed to R.

It is evident that the rudimentary functionality of the standard R GUI makes R unattractive for those users familiar with commercial statistical packages characterized by their friendly interfaces. As improved substitutes for the R GUI environment and its script editor, several applications are offered to facilitate the access to menus and make the task of writing R scripts easier. Among these projects, the two applications RStudio (RStudio Team 2015) and TinnR (Faria et al. 2015) stand out. R is case sensitive, and R programming requires the frequent use of parentheses and brackets, and so on. TinnR and RStudio make use of colors to indicate corresponding opening and closing parentheses, brackets, and so on, as well as to differentiate between sentences from the R language, arguments of functions, and comments. None of these facilities are available in the script editor of the R GUI. You should therefore probably choose one of these friendlier applications to improve your interaction with R.

A.2   Help files in R

R includes a help system, which is useful in providing basic information about each R command, such as its syntax, the outcome/object produced, examples of use, and references from where the command is based. It is essential to rely on this standard help system whenever there is doubt about the correct syntax of R statements. In R GUI, you can access the help system as one option of the menu bar, choosing different search strategies. Actually, the help system provides more information than just command syntax. You can also access R documentation and manuals, learn about the development of R, and get answers to frequently asked questions.

A.3   R packages

A *package* in R is a set of related/integrated R functions, help files, and data sets that either the user or R itself invokes and makes available for a particular purpose. With the exception of a group of packages already installed with R, the user will need to download and install a package of interest into R. Installation is only carried out once, and each time a package of this sort is required, the user only needs to load it into the R environment for the current session. The main public package repository is available in the CRAN mirror sites all over the world, but it is wise to choose the closest location to download and install a package. Accessing the Packages menu in R GUI is the best way to select repositories and mirror sites and to run the automatic installation of packages and their so-called dependencies, that is, additional packages necessary for the functionality of the one in which you are interested. A package is actually a zip file that is saved in an internal or external storage medium, and it contains the necessary procedures, documentation, and connectivity to R. It is also possible to install a package locally (an option present in the Packages menu), but its functionality might be affected if any dependency is absent in the set of installed packages. R offers a huge variety of packages. For example, in Version 3.1.3, the CRAN package repository features 6415 available packages. However, in this book, we will just indicate the packages that are necessary to get the results for the examples presented.

A.4   Objects in R

The R language manipulates *objects*. These are defined as entities that can be represented on a computer: numbers, variables, matrices, user functions, data sets (known as *data frames* in R), or a combination of all these (such as lists). A name can be given to any object, as long as the sequence of characters forming the name does not contain any of the following characters:

space(s), $-$, $+$, $*$, $/$, #, %, &, [,], {, }, (,), or ~. In addition, object names cannot start with a number, and they are case sensitive. Thus, X and x are different and refer to different objects. To avoid subsequent typographical errors, it is recommended that you use short and mnemonic object names. Assignment of an object name means that the object is allocated to the current workspace of R via the assignment operator, composed by the character < and a hyphen, with no space between them, i.e. <-. As an example, you may want to define a variable called S carrying the value 12. The command is then S <- 12.

   If S does not exist, it will be created. Otherwise, its previous contents are replaced. Then, S will be kept in the workspace (in the memory of the computer), but other commands are needed (e.g., *save*) to store it to a disk. There are also ways to get rid of an object in the current session through the command rm.

A.5   Vectors in R

The main object in R is the vector. It is an n-tuple of objects of the same class. The familiar vector of real numbers is the best example of a vector in R that can be created using the concatenation function c(). As an example, consider the data shown in Table 1.3. Assume that you would like to analyze the annual precipitation of the sites where colonies of the butterfly *Euphydryas editha* were sampled. To do this, you would ask R to store the values in the numeric vector

```
precip <- c(43, 20, 28, 28, 28, 15, 21, 10, 10, 19, 22,
    58, 34, 21, 42, 50)
```

   Here, precip is a vector, a single-row array whose length is equal to its number of elements, in this case, 16. You can verify this length by writing

```
length(precip)
[1] 16
```

where the second line gives the response to the first line.

   Each element in a vector has its own index or column number, and this can be referred to by enclosing it in square brackets. As an example, the 13th element of precip is precipitation 34 mm. You can display this particular value in R as

```
precip[13]
[1] 34
```

   Names can be given to each element of a numeric vector such as precip. These names can be taken from a character vector (i.e. a vector containing alphanumeric characters enclosed in quotation marks). As an example, to assign names to the elements of precip using the colony abbreviations as identifiers, the next three commands can be used:

```
COLONY <- c("SS", "SB", "WSB", "JRC", "JRH", "SJ", "CR",
   + "UO", "LO", "DP", "PZ", "MC", "IF", "AF", "GH", "GL")
names(precip) <- COLONY
precip
SS SB WSB JRC JRH SJ CR UO LO DP PZ MC IF AF GH GL
43 20  28  28  28 15 21 10 10 19 22 58 34 21 42 50
```

The most important feature of vectors is that they can only hold one class of object (either numbers, strings of characters, levels of one factor, or logical values) but not a mixture of them. In addition, vectors can be operated on using arithmetic expressions applied element by element. Refer to Chapter 2 in the manual *Introduction to R* (Venables et al., 2015) for more examples and further properties of vectors.

A.6   Matrices in R

Matrices are particular cases of what are known in R as arrays of data of the same class, which is the multidimensional generalization of vectors with multiples. There are several ways to build up a matrix. One possibility is to convert it from a vector using the matrix function. As an example, assume that we have an eight-dimensional vector containing the number of species (species richness) in four sites, with the first four elements of the vector being the richness values for the four sites in Year 1 and the last four elements the corresponding richness values for Year 2, as follows:

```
RICHNESS <- c(2, 2, 3, 2, 2, 1, 5, 1)
```

A better way to arrange these data is by defining a 4 × 2 matrix with rows indicating the sites and columns corresponding to years:

```
RICHMAT <- matrix(RICHNESS, nrow = 4)
RICHMAT
     [,1]  [,2]
[1,]   2    2
[2,]   2    1
[3,]   3    5
[4,]   2    1
```

By default, a matrix is filled columnwise. Here, it was only necessary to write the argument nrow = 4 to indicate the number of rows, but you can be more specific by writing the number of rows (nrow=) and the number of columns (ncol=) as

```
RICHMAT <- matrix(RICHNESS, nrow = 4, ncol=2).
```

In this example, R addresses each element of the matrix RICHMAT by its corresponding row and column position. Thus, the observed richness for Site 3 on Year 2 can be found as

```
RICHMAT[3,2]
[1] 5
```

The elements of a particular row or column can be referred to by omitting the corresponding column and row, after or before the comma, respectively. Thus, the richness for Years 1 and 2 in Site 2 is

```
RICHMAT[2,]
[1] 2 1
```

Similarly, the observed richness in Year 1 for all the sites is

```
> RICHMAT[,1]
[1] 2 2 3 2
```

It is also possible to build a matrix by combining two or more vectors of the same length (and class). Following a similar idea from the previous example, assume that there are two vectors of four entries each, with the first vector corresponding to the richness found in four sites for Year 1 and the second one containing the corresponding richness for Year 2, defined as

```
richy1 <- c(2, 2, 3, 2)
richy2 <- c(2, 1, 5, 1)
```

The vectors can then be combined (bound) in one matrix using cbind() (combine by columns) or rbind() (combine by rows). Then, combining by columns is given by

```
RICHMAT1 <- cbind(richy1, richy2)
```

so that

```
RICHMAT1
richy1 richy2
[1,]  2    2
[2,]  2    1
[3,]  3    5
[4,]  2    1
```

In this case, R assigns column names to the matrix RICHMAT1 taken from the source vectors richy1 and richy2. The following code produces the same output (namely, the first column of RICHMAT1):

```
RICHMAT1[,1]
[1] 2 2 3 2
RICHMAT1[,"richy1"]
[1] 2 2 3 2
```

Column names in a matrix can be invoked or defined by means of the command colnames. For example, to check whether the column names for RICHMAT1 are already in use, try

```
colnames(RICHMAT1)
[1] "richy1" "richy2"
```

Similarly, row names of a matrix can be displayed or defined using the command rownames. When RICHMAT1 was created, it did not have row names. This can be checked by

```
rownames(RICHMAT1)
NULL
```

Suitable row names can then be given to RICHMAT1:

```
rownames(RICHMAT1) <- c("S1","S2","S3","S4")
RICHMAT1
richy1  richy2
S1   2   2
S2   2   1
S3   3   5
S4   2   1
```

Now, the rows of RICHMAT1 correspond to each of four sites, identified by the letter S and the site number. Additional properties of arrays and matrices can be found in Venables et al. 2015 (chapter 5) and in the Appendices of Chapters 2 and 3.

A.7   *Arranging multivariate data in R from source data bases (data frames)*

To manipulate multivariate data and handle different sorts of variables, R has a special class of object called a *data frame*. Data frames and two-dimensional arrays (matrices) have one property in common: the data are arranged in rows and columns. However, the rows of a data frame are identified as different sampling units from where observations or measurements are made, with these being placed in different columns, so that each column in a data frame corresponds to a particular variable. Therefore, unlike arrays that only allow one class of data, data frames may include a mixture of variables: numeric, character, factor, logical, dates, and so on.

One explicit way of constructing a data frame is to combine different vectors, presumably of the same length and each one containing a particular variable, and declare them in the data.frame command:

```
my.data <- data.frame(vector1, vector2, ...)
```

A better practical way of making a data frame takes advantage of R being capable of reading data files in different formats, such as tab- or comma-separated values (*csv*), which is the most convenient format for importing data into R. Alternatively, data saved in a spreadsheet and statistical applications such as Excel, SAS, or Minitab can be imported, in which case it is necessary to install suitable R packages to read the data.

Here, we will only consider the making of a data frame using the command `read.table` applied to an external file saved with tab-separated values or csv. As an example, consider the data described in Example 1.5. These data can be accommodated in a data frame with 30 rows (European countries) and 10 variables, with nine of the variables corresponding to the percentages (i.e., nine numeric variables) of the labor force in nine different types of industry for the European countries considered, and one categorical variable referring to the political grouping (the EU, the EFTA, the Eastern European countries, etc.).

Assume that the data on the percentages of people employed in different industry groups in Europe have been saved as a tab-delimited file with the name *Euroemp.txt*, where this file can be found at the website http://www.manly-biostatistics.co.nz/. The first two and last two lines and headings of each column of the data are then

Country	Group	AGR	MIN	MAN	PS	CON	SER	FIN	SPS	TC
Belgium	EU	2.6	0.2	20.8	0.8	6.3	16.9	8.7	36.9	6.8
Denmark	EU	5.6	0.1	20.4	0.7	6.4	14.5	9.1	36.3	7.0
...	...	...	...	...	...	...	...	...	...	...
Malta	Other	2.6	0.6	27.9	1.5	4.6	10.2	3.9	41.6	7.2
Turkey	Other	44.8	0.9	15.3	0.2	5.2	12.4	2.4	14.5	4.4

Here, Country is not a variable; it refers to row names. This can be taken into account when importing the data using read.table. The second column is a qualitative variable (Group), and the last nine columns are all numeric variables corresponding to the percentage of people employed to a different industry group. The command needed to produce a data frame from the tab-delimited file *Euroemp.txt* is then

```
euro.emp <- read.table("Euroemp.txt", header=TRUE, row.
    names=1)
```

This command assumes that a working directory has been chosen in R such that the directory (folder) contains the file *Euroemp.txt*. One way to accomplish this is by accessing the menu **File > Change dir ...** in R GUI. The first argument is the file name enclosed in quotation marks, followed by the option header =TRUE, which means that the file contains a header (names assigned to each column). However, the first column is not a variable; it refers to row (sampling unit) names. This explains the option row. names=1, meaning *the row names are located in column 1*. By default, R tries to read a tab- or space-delimited source file and then to convert it into a data frame. In the example, the data frame euro.emp was created. This contains the same information as the original source file, but it is now accessible to R. To confirm that this is indeed a data frame, you can type

```
class(euro.emp)
[1] "data.frame"
```

You can then display the whole data frame in the console by typing

```
euro.emp
```

or the first six rows:

```
head(euro.emp)
```

One helpful way to recognize the type of variables present in a data frame is through the str command (i.e., the data frame *structure*). In this case,

```
str(euro.emp)
'data.frame':   30 obs. of 10 variables:
$ Group: Factor w/ 4 levels "Eastern", "EFTA",..: 3 3 3
   3 3 3 3 3 3 3 ...
$ AGR  : num 2.6 5.6 5.1 3.2 22.2 13.8 8.4 3.3 4.2 11.5 ...
$ MIN  : num 0.2 0.1 0.3 0.7 0.5 0.6 1.1 0.1 0.1 0.5 ...
$ MAN  : num 20.8 20.4 20.2 24.8 19.2 19.8 21.9 19.6 19.2
   23.6 ...
$ PS   : num 0.8 0.7 0.9 1 1 1.2 0 0.7 0.7 0.7 ...
$ CON  : num 6.3 6.4 7.1 9.4 6.8 7.1 9.1 9.9 0.6 8.2 ...
$ SER  : num 16.9 14.5 16.7 17.2 18.2 17.8 21.6 21.2 18.5
   19.8 ...
$ FIN  : num 8.7 9.1 10.2 9.6 5.3 8.4 4.6 8.7 11.5 6.3 ...
$ SPS  : num 36.9 36.3 33.1 28.4 19.8 25.5 28 29.6 38.3
   24.6 ...
$ TC   : num 6.8 7 6.4 5.6 6.9 5.8 5.3 6.8 6.8 4.8 ...
```

Here, the imported variables are listed in different lines, each starting with a dollar sign. This pinpoints the manner in which each variable contained in a data frame is referred to in R. The name of the variable is preceded by the name of the data frame, and these two are separated by the $ character. Also, to display the variable (vector) AGR in the data frame euro.emp, the command is

```
euro.emp$AGR
 [1]  2.6 5.6 5.1 3.2 22.2 13.8 8.4 3.3 4.2 11.5 9.9 2.2 7.4 8.5
[15] 10.5 5.8 3.2 5.6 55.5 19.0 12.8 15.3 23.6 22.0 18.5 5.0
   13.5 0.0
[29]  2.6 44.8
```

An error message is produced if you request a variable without a data frame name, so that this gives

```
AGR
Error: object 'AGR' not found
```

This rule of combining the data frame and the name of a variable can be avoided with suitable code executed before calling the variable, for example using the commands attach or with (see Venables et al. 2015, Chapter 6). It is important to bear in mind that a data frame is a special class of object in R, known as *list*, which is considered one of the most versatile objects in R for handling data or the results produced by other R functions. See Venables et al. (2015) for more information.

By default, R converts (coerces) any variable containing alphanumeric characters into a *factor*. This is euro.emp$Group in the above example. A factor is an object (a vector) useful to group the components of other vectors of the same length. The main difference between a factor vector and a character vector is that the components of the former are not enclosed in quotation marks. Internally, R handles a factor as a numeric (integer) vector whose order is dictated, by default, by the alphabetic order of the labels. In terms of computer memory usage, storage of factors is better, because an integer uses fewer bytes of memory than a string of characters. These properties can be seen by specific R commands. Thus, the vector euro.emp$Group is of factor class, as shown by

```
class(euro.emp$Group)
[1] "factor"
```

but its mode is numeric:

```
mode(euro.emp$Group)
[1] "numeric"
```

Regarding the example of nine different types of industry in European countries, the data could also be stored as a csv file. The R command that can be used for making a data frame containing these data would then be

```
euro.empcsv <- read.table("Euroemp.csv", header=TRUE,
    row.names=1, sep=",")
```

Here, the last argument sep="," indicates a comma as the separator of each field in the dataset. Alternatively, a csv file can be read instead by the command read.csv:

```
euro.empcsv <- read.csv("Euroemp.csv", header=TRUE,
    row.names=1)
```

Here, read.csv is identical to read.table, except that a comma is the default separator for the file to be imported. See Chapter 7 in Venables et al. (2015) for further topics related to reading data from files.

A.8 Data frame indexing (single and double)

Data frames enjoy the properties of single and double indexing described above for vectors and matrices. Thus, specific values in an individual vector (variable) within a data frame can be accessed by referring to only one index (indicating the position of the datum or the data for that variable), whereas the access of groups of variables and/or groups of rows is possible through double indexing (of rows and columns).

The contents of the variable AGR in the data frame euro.emp can be verified as

```
euro.emp$AGR[3:8]
[1]  5.1  3.2  22.2  13.8  8.4  3.3
```

where 3:8 is R shorthand for the vector c(3, 4, 5, 6, 7, 8). Similarly,

```
euro.emp$AGR[c(1,3,5)]
[1]  2.6  5.1  22.2
```

Some examples of double indexing are

```
euro.emp[25:30,]
Group AGR MIN MAN PS CON SER FIN SPS TC
USSRF         Eastern 18.5 0.0 28.8 0.0 10.2 7.9 0.6 25.6
              8.4
YugoslaviaF Eastern 5.0 2.2 38.7 2.2 8.1 13.8 3.1 19.1 7.8
Cyprus        Other 13.5 0.3 19.0 0.5 9.1 23.7 6.7 21.2 6.0
Gibraltar     Other 0.0 0.0 6.8 2.0 16.9 24.5 10.8 34.0 5.0
Malta         Other 2.6 0.6 27.9 1.5 4.6 10.2 3.9 41.6 7.2
Turkey        Other 44.8 0.9 15.3 0.2 5.2 12.4 2.4 14.5 4.4
```

and

```
euro.emp[c("UK","Romania"),]
Group AGR MIN MAN PS CON SER FIN SPS TC
UK      EU 2.2 0.7 21.3 1.2 7.0 20.2 12.4 28.4 6.5
Romania Eastern 22.0 2.6 37.9 2.0 5.8 6.9 0.6 15.3 6.8
```

where particular rows are referred by their name, and

```
euro.emp[,c(2,5)]
AGR    PS
Belgium    2.6      0.8
  ...              ...        ...
Turkey     44.8     0.2
```

which is asking for the second and the fifth variables (for the sake of space only the first and the last rows are shown). Also,

```
euro.emp[,c("Group",  "MIN")]
Group      MIN
Belgium    EU     0.2
...        ...    ...
Turkey     Other  0.9
```

is asking for the contents of variables "Group" and "MIN," with only the first and last lines shown here. The command

```
euro.emp[euro.emp$Group == "Other",]
Group AGR MIN MAN PS CON SER FIN SPS TC
Cyprus    Other 13.5 0.3 19.0 0.5 9.1 23.7 6.7 21.2 6.0
Gibraltar Other 0.0 0.0 6.8 2.0 16.9 24.5 10.8 34.0 5.0
Malta     Other 2.6 0.6 27.9 1.5 4.6 10.2 3.9 41.6 7.2
Turkey    Other 44.8 0.9 15.3 0.2 5.2 12.4 2.4 14.5 4.4
```

is only the sampling units (rows) fulfilling the condition that the variable Group is equal to "Other" which is asking for the second and the fifth variables (for the sake of space only the first and the last rows are shown). The double equals sign == refers to a conditional or logical equality sign, so that the rows shown are only those for which it is TRUE that the Group is equal to "Other."

For further examples of the use of single or double indexing in data frames, see Logan (2010, chapter 2).

References

Faria, J.C., Grosjean, P., and Jelihovschi, E. (2015). *Tinn-R Editor - GUI for R Language and Environment.* http://nbcgib.uesc.br/lec/software/editores/tinn-r/en

RStudio Team. (2015). *RStudio: Integrated Development for R.* RStudio, Inc., Boston, MA. http://www.rstudio.com/

chapter two

Matrix algebra

2.1   The need for matrix algebra

The theory of multivariate statistical methods can only be explained reasonably well with the use of matrix algebra. For this reason, it is helpful, if not essential, to have at least some knowledge of this area of mathematics. This is true even for those who are only interested in using the methods as tools. At first sight, the notation of matrix algebra is certainly somewhat daunting. However, it is not difficult to understand the basic principles, providing that some of the details are accepted on faith.

2.2   Matrices and vectors

An m × n *matrix* is an array of numbers with m rows and n columns, considered as a single entity, of the form

$$
\mathbf{A} = \begin{bmatrix}
a_{11} & \cdots & a_{12} & \cdots & a_{1n} \\
a_{21} & \cdots & a_{22} & \cdots & a_{2n} \\
\cdot & & \cdot & & \cdot \\
\cdot & & \cdot & & \cdot \\
a_{m1} & a_{m2} & \cdots & & a_{mn}
\end{bmatrix}
$$

If m = n, then it is a *square* matrix. If there is only one column, such as

$$
\mathbf{c} = \begin{bmatrix}
c_1 \\
c_2 \\
\cdot \\
\cdot \\
c_m
\end{bmatrix}
$$

then this is called a *column vector*. If there is only one row, such as

$$
\mathbf{r} = \left(r_1, r_2, \dots r_n \right)
$$

then this is called a *row vector*. Bold type is used to indicate matrices and vectors.

The *transpose* of a matrix is obtained by interchanging the rows and the columns. Thus, the transpose of the matrix $\mathbf{A}$ is

$$
\mathbf{A}' = \begin{bmatrix}
a_{11} & \cdot\cdot & a_{12} & \cdot\cdot & a_{m1} \\
a_{12} & \cdot\cdot & a_{22} & \cdot\cdot & a_{m2} \\
\cdot & & \cdot & & \cdot \\
\cdot & & \cdot & & \cdot \\
a_{1n} & \cdot\cdot & a_{2n} & \cdot\cdot & a_{mn}
\end{bmatrix}
$$

Also, the transpose of the vector $\mathbf{c}$ is $\mathbf{c}' = (c_1, c_2, \ldots, c_m)$, and the transpose of the row vector $\mathbf{r}$ is the column vector $\mathbf{r}'$.

There are a number of special kinds of matrices that are important. A *zero matrix* has all elements equal to zero, so that it is of the form

$$
\mathbf{0} = \begin{bmatrix}
0 & \cdot\cdot & 0 & \cdot\cdot & 0 \\
0 & \cdot\cdot & 0 & \cdot\cdot & 0 \\
\cdot & & \cdot & & \cdot \\
\cdot & & \cdot & & \cdot \\
0 & \cdot\cdot & 0 & \cdot\cdot & 0
\end{bmatrix}
$$

A *diagonal matrix* has zero elements except down the main diagonal, so that it takes the form

$$
\mathbf{D} = \begin{bmatrix}
d_1 & \cdot\cdot & 0 & \cdot\cdot & 0 \\
0 & \cdot\cdot & d_2 & \cdot\cdot & 0 \\
\cdot & & \cdot & & \cdot \\
\cdot & & \cdot & & \cdot \\
0 & \cdot\cdot & 0 & \cdot\cdot & d_n
\end{bmatrix}
$$

A *symmetric matrix* is a square matrix that is unchanged when it is transposed, so that $\mathbf{A}' = \mathbf{A}$. Finally, an *identity matrix* is a diagonal matrix with all terms on the diagonal equal to one, so that

$$I = \begin{bmatrix} 1 & \cdot\cdot & 0 & \cdot\cdot & 0 \\ 0 & \cdot\cdot & 1 & \cdot\cdot & 0 \\ \cdot & & \cdot & & \cdot \\ \cdot & & \cdot & & \cdot \\ 0 & \cdot\cdot & 0 & \cdot\cdot & 1 \end{bmatrix}$$

Two matrices are equal only if they are the same size and all their elements are equal. For example,

$$\begin{bmatrix} a_{11} & \cdot\cdot & a_{12} & \cdot\cdot & a_{13} \\ a_{21} & \cdot\cdot & a_{22} & \cdot\cdot & a_{23} \\ a_{31} & \cdot\cdot & a_{32} & \cdot\cdot & a_{33} \end{bmatrix} = \begin{bmatrix} b_{11} & \cdot\cdot & b_{12} & \cdot\cdot & b_{13} \\ b_{21} & \cdot\cdot & b_{22} & \cdot\cdot & b_{23} \\ b_{31} & \cdot\cdot & b_{32} & \cdot\cdot & b_{33} \end{bmatrix}$$

only if $a_{11} = b_{11}$, $a_{12} = b_{12}$, $a_{13} = b_{13}$, and so on.

The *trace* of a matrix is the sum of the diagonal terms, which is only defined for a square matrix. For example, the trace of the 3×3 matrix with the elements a_{ij} shown above has trace(A) = $a_{11} + a_{22} + a_{33}$.

2.3   Operations on matrices

The ordinary arithmetic processes of addition, subtraction, multiplication, and division have their counterparts with matrices. With addition and subtraction, it is just a matter of working element by element with two matrices of the same size. For example, if A and B are both of size 3×2, then

$$\mathbf{A} + \mathbf{B} = \begin{bmatrix} a_{11} & \cdot\cdot & a_{12} \\ a_{21} & \cdot\cdot & a_{22} \\ a_{31} & \cdot\cdot & a_{32} \end{bmatrix} + \begin{bmatrix} b_{11} & \cdot\cdot & b_{12} \\ b_{21} & \cdot\cdot & b_{22} \\ b_{31} & \cdot\cdot & b_{32} \end{bmatrix} = \begin{bmatrix} a_{11} + b_{11} & \cdot\cdot & a_{12} + b_{12} \\ a_{21} + b_{21} & \cdot\cdot & a_{22} + b_{22} \\ a_{31} + b_{31} & \cdot\cdot & a_{32} + b_{32} \end{bmatrix}$$

while

$$\mathbf{A} - \mathbf{B} = \begin{bmatrix} a_{11} & \cdot\cdot & a_{12} \\ a_{21} & \cdot\cdot & a_{22} \\ a_{31} & \cdot\cdot & a_{32} \end{bmatrix} + \begin{bmatrix} b_{11} & \cdot\cdot & b_{12} \\ b_{21} & \cdot\cdot & b_{22} \\ b_{31} & \cdot\cdot & b_{32} \end{bmatrix} = \begin{bmatrix} a_{11} - b_{11} & \cdot\cdot & a_{12} - b_{12} \\ a_{21} - b_{21} & \cdot\cdot & a_{22} - b_{22} \\ a_{31} - b_{31} & \cdot\cdot & a_{32} - b_{32} \end{bmatrix}$$

Clearly, these operations only apply with two matrices of the same size.

In matrix algebra, an ordinary number such as 20 is called a *scalar.* Multiplication of a matrix **A** by a scalar k is then defined to be the multiplication of every element in **A** by k. Thus, if **A** is the 3 × 2 matrix as shown above, then

$$k\mathbf{A} = \begin{bmatrix} ka_{11} & \cdot\cdot & ka_{12} \\ ka_{21} & \cdot\cdot & ka_{22} \\ ka_{31} & \cdot\cdot & ka_{32} \end{bmatrix}$$

The multiplication of two matrices, denoted by **A.B** or **A** × **B**, is more complicated. To begin with, **A.B** is only defined if the number of columns of **A** is equal to the number of rows of **B**. Assume that this is the case, with **A** having the size m × n and **B** having the size n × p. Then, multiplication is defined to produce the result

$$\mathbf{A} \cdot \mathbf{B} = \begin{bmatrix} a_{11} & \cdot\cdot & a_{12} & \cdot\cdot & a_{1n} \\ a_{21} & \cdot\cdot & a_{22} & \cdot\cdot & a_{2n} \\ \cdot & & \cdot & & \cdot \\ \cdot & & \cdot & & \cdot \\ a_{m1} & \cdot\cdot & a_{m2} & \cdot\cdot & a_{mn} \end{bmatrix} \begin{bmatrix} b_{11} & \cdot\cdot & b_{12} & \cdot\cdot & b_{1p} \\ b_{21} & \cdot\cdot & b_{22} & \cdot\cdot & b_{2p} \\ \cdot & & \cdot & & \cdot \\ \cdot & & \cdot & & \cdot \\ b_{n1} & \cdot\cdot & b_{n2} & \cdot\cdot & b_{np} \end{bmatrix}$$

$$= \mathbf{A}' = \begin{bmatrix} \Sigma a_{1j} \cdot b_{j1} & \cdot\cdot & \Sigma a_{1j} \cdot b_{j2} & \cdot\cdot & \Sigma a_{1j} \cdot b_{jp} \\ \Sigma a_{2j} \cdot b_{j1} & \cdot\cdot & \Sigma a_{2j} \cdot b_{j2} & \cdot\cdot & \Sigma a_{2j} \cdot b_{jp} \\ \cdot & & \cdot & & \cdot \\ \cdot & & \cdot & & \cdot \\ \Sigma a_{mj} \cdot b_{j1} & \cdot\cdot & \Sigma a_{mj} \cdot b_{j2} & \cdot\cdot & \Sigma a_{mj} \cdot b_{jp} \end{bmatrix}$$

where the summations are for j from 1 to n. Hence, the element in the ith row and jth column of **A.B** is

$$\Sigma a_{ij} \cdot b_{jk} = a_{i1} \cdot b_{1k} + a_{i2} \cdot b_{2k} + \ldots + a_{in} b_{nk}$$

When A and B are both square matrices, then **A.B** and **B.A** are both defined. However, they are not generally equal. For example,

$$\begin{bmatrix} 2 & -1 \\ 1 & 1 \end{bmatrix}\begin{bmatrix} 1 & 1 \\ 0 & 1 \end{bmatrix} = \begin{bmatrix} 2\times1-1\times0 & 2\times1-1\times1 \\ 1\times1+1\times0 & 1\times1+1\times1 \end{bmatrix} = \begin{bmatrix} 2 & 1 \\ 1 & 2 \end{bmatrix}$$

whereas

$$\begin{bmatrix} 1 & 1 \\ 0 & 1 \end{bmatrix}\begin{bmatrix} 2 & -1 \\ 1 & 1 \end{bmatrix} = \begin{bmatrix} 1\times2+1\times1 & -1\times1+1\times1 \\ 0\times2+1\times1 & -1\times0+1\times1 \end{bmatrix} = \begin{bmatrix} 3 & 0 \\ 1 & 1 \end{bmatrix}$$

2.4 Matrix inversion

Matrix inversion is analogous to the ordinary arithmetic process of division. For a scalar k, it is of course true that $k \times k^{-1} = 1$. In a similar way, if **A** is a square matrix and

$$\mathbf{A} \times \mathbf{A}^{-1} = \mathbf{I},$$

where **I** is the identity matrix, then the matrix $\mathbf{A}^{-1}$ is the *inverse* of the matrix **A**. Inverses only exist for square matrices, but not all square matrices have inverses. If an inverse does exist, then it is both a left inverse, so that $\mathbf{A}^{-1} \times \mathbf{A} = \mathbf{I}$, as well as a right inverse, so that $\mathbf{A} \times \mathbf{A}^{-1} = \mathbf{I}$.

An example of an inverse matrix is

$$\begin{bmatrix} 2 & 1 \\ 1 & 2 \end{bmatrix}^{-1} = \begin{bmatrix} 2/3 & -1/3 \\ -1/3 & 2/3 \end{bmatrix}$$

which can be verified by checking that

$$\begin{bmatrix} 2 & 1 \\ 1 & 2 \end{bmatrix}\begin{bmatrix} 2/3 & -1/3 \\ -1/3 & 2/3 \end{bmatrix} = \begin{bmatrix} 1 & 0 \\ 0 & 1 \end{bmatrix}$$

Actually, the inverse of a 2×2 matrix, if it exists, can be calculated fairly easily. The equation is

$$\begin{bmatrix} a & b \\ c & d \end{bmatrix}^{-1} = \begin{bmatrix} d/\Delta & -b/\Delta \\ -c/\Delta & a/\Delta \end{bmatrix}$$

where $\Delta = a \times d - b \times c$. Here, the scalar Δ is called the *determinant* of the matrix being inverted. Clearly, the inverse is not defined if $\Delta = 0$, because finding the elements of the inverse then involves a division by zero. For 3×3 and larger matrices, the calculation of the inverse is tedious and best done by using a computer program. Nowadays, even spreadsheets include a facility to compute an inverse.

Any square matrix has a determinant, which can be calculated by a generalization of the equation just given for the 2×2 case. If the determinant is zero, then the inverse does not exist, and vice versa. A matrix with a zero determinant is said to be *singular*.

Matrices sometimes arise for which the inverse is equal to the transpose. They are then said to be *orthogonal*. Hence, A is orthogonal if $\mathbf{A}^{-1} = \mathbf{A}'$.

2.5   Quadratic forms

Suppose that $\mathbf{A}$ is an n by n matrix and $\mathbf{x}$ is a column vector of length n. Then, the quantity

$$Q = \mathbf{x}'\mathbf{A}\mathbf{x}$$

is a scalar that is called a *quadratic form*. This scalar can also be expressed as

$$Q = \sum_{i=1}^{n} \sum_{j=1}^{n} x_i a_{ij} x_j$$

where x_i is the element in the ith row of $\mathbf{x}$, and a_{ij} is the element in the ith row and jth column of $\mathbf{A}$.

2.6   Eigenvalues and eigenvectors

Consider the set of linear equations

$$a_{11}x_1 + a_{12}x_2 + \ldots + a_{1n}x_n = \lambda\, x_1$$

$$a_{21}x_1 + a_{22}x_2 + \ldots + a_{2n}x_n = \lambda\, x_2$$

$$a_{n1}x_1 + a_{n2}x_2 + \ldots + a_{nn}x_n = \lambda\, x_n$$

where λ is a scalar. These can also be written in matrix form as

$$\mathbf{A}\,\mathbf{x} = \lambda\mathbf{x}$$

or

$$(\mathbf{A} - \lambda\mathbf{I})\mathbf{x} = 0,$$

where $\mathbf{I}$ is the $n \times n$ identity matrix and $\mathbf{0}$ is an $n \times 1$ vector of zeros. Then, it can be shown that these equations can only hold for certain particular values of λ, which are called the *latent roots* or *eigenvalues* of $\mathbf{A}$. There can be up to n of these eigenvalues. Given the ith eigenvalue λ_i, the equations can be solved by arbitrarily setting $x_1 = 1$, and the resulting vector of x values with transpose $\mathbf{x'} = (1, x_2, x_3, \dots, x_n)$, or any multiple of this vector, is called the ith *latent root* or the ith *eigenvector* of the matrix $\mathbf{A}$. Also, the sum of the eigenvalues is equal to the trace of A defined in Section 2.2, so that $\text{trace}(\mathbf{A}) = \lambda_1 + \lambda_2 + \dots + \lambda_n$.

2.7 Vectors of means and covariance matrices

Population and sample values for a single random variable are often summarized by the values for the mean and variance. Thus, if a sample of size n yields the values $x_1, x_2, \dots, x_n$, then the *sample mean* is defined to be

$$\bar{x} = x_1 + x_2 + \dots + x_n = \sum_{i=1}^{n} x_i / n$$

while the *sample variance* is

$$s^2 = \sum_{i=1}^{n} (x_i - \bar{x})^2 / (n-1)$$

These are estimates of the corresponding population parameters, which are the *population mean* μ and the *population variance* σ^2.

In a similar way, multivariate populations and samples can be summarized by *mean vectors* and *covariance matrices*. Suppose that there are p variables $X_1, X_2, \dots X_p$ being considered, and that a sample of n values for each of these variables is available. Let the sample mean and sample variance for the jth variable be $\bar{x}_j$ and s_j^2, respectively, where these are calculated using the equations given in the previous paragraph. In addition, define the *sample covariance* between variables X_j and X_k by

$$c_{jk} = \sum_{i=1}^{n} (x_{ij} - \bar{x}_j)(x_{ik} - \bar{x}_k) / (n-1)$$

where x_{ij} is the value of variable X_j for the ith multivariate observation. This covariance is then a measure of the extent to which there is a linear relationship between X_j and X_k, with a positive value indicating that large values of X_j and X_k tend to occur together, and a negative value indicating that large values for one variable tend to occur with small values for the other variable. It is related to the ordinary correlation coefficient between the two variables, which is defined to be

$$r_{jk} = c_{jk} / (s_j s_k)$$

Furthermore, the definitions imply that $c_{kj} = c_{jk}$, $r_{kj} = r_{jk}$, $c_{jj} = s_j^2$, and $r_{jj} = 1$. With these definitions, the transpose of the *sample mean vector* is

$$\mathbf{x}' = (x_1, x_2, \ldots, x_p),$$

which can be thought of as reflecting the center of the multivariate sample. It is also an estimate of the transpose of the population vector of means

$$\boldsymbol{\mu}' = (\mu_1, \mu_2, \ldots, \mu_p)$$

Furthermore, the sample matrix of variances and covariances, or the *covariance matrix*, is

$$
\mathbf{C} = \begin{bmatrix}
c_{11} & \cdots & c_{12} & \cdots & c_{1p} \\
c_{21} & \cdots & c_{22} & \cdots & c_{2p} \\
\cdot & & \cdot & & \cdot \\
\cdot & & \cdot & & \cdot \\
c_{p1} & \cdots & c_{p2} & \cdots & c_{pp}
\end{bmatrix}
$$

where $c_{ii} = s_i^2$. This is also sometimes called the *sample dispersion matrix*, and it measures the amount of variation in the sample as well as the extent to which the p variables are correlated. It is an estimate of the *population covariance matrix*

$$
\boldsymbol{\Sigma} = \begin{bmatrix}
\sigma_{11} & \cdots & \sigma_{12} & \cdots & \sigma_{1p} \\
\sigma_{21} & \cdots & \sigma_{22} & \cdots & \sigma_{2p} \\
\cdot & & \cdot & & \cdot \\
\cdot & & \cdot & & \cdot \\
\sigma_{p1} & \cdots & \sigma_{p2} & \cdots & \sigma_{pp}
\end{bmatrix}
$$

Finally, the *sample correlation matrix* is

$$
R = \begin{bmatrix}
1 & \cdot\cdot & r_{12} & \cdot\cdot & r_{1p} \\
r_{21} & \cdot\cdot & 1 & \cdot\cdot & r_{2p} \\
\cdot & & \cdot & & \cdot \\
\cdot & & \cdot & & \cdot \\
r_{p1} & \cdot\cdot & r_{p2} & \cdot\cdot & 1
\end{bmatrix}
$$

Again, this is an estimate of the corresponding *population correlation matrix*. An important result for some analyses is that if the observations for each of the variables are coded by subtracting the sample mean and dividing by the sample standard deviation, then the coded values will have a mean of zero and a standard deviation of one for each variable. In that case, the sample covariance matrix will equal the sample correlation matrix, that is, $C = R$.

2.8   Further reading

This short introduction to matrix algebra will suffice for understanding the methods described in the remainder of this book and some of the theory behind these methods. However, for a better understanding of the theory, more knowledge and proficiency are required.

There are many books of various lengths that cover what is needed just for statistical applications. Four of these are by Searle (2006), Healy (2000), Harville (2000), and Namboodiri (1984). Another possibility is to do a web search on the topic of matrix algebra. This should yield some useful free books and course notes.

References

Harville, D.A. (2000). *Matrix Algebra from a Statistician's Perspective.* New York: Springer.
Healy, M.J.R. (2000). *Matrices for Statistics.* 2nd Edn. Oxford: Clarendon.
Namboodiri, K. (1984). *Matrix Algebra: An Introduction.* Thousand Oaks, CA: Sage.
Searle, S.R. (2006). *Matrix Algebra Useful to Statisticians.* New York: Wiley-Interscience.

Appendix: Matrix Algebra in R

A summary of the main R functions useful for handling matrices and performing matrix operations is given here. The complete set of options for each function and alternative functions can be found in the corresponding R help documents. A more detailed description and illustrative examples are given in the website for this book. Here, **x** is a vector, and **A and B** are matrices.

Function(s)	Description	Some useful options
`matrix(x, nrow=, ncol=,...)`	Generates a matrix columnwise with nrow rows and ncol columns	`byrow=TRUE` generates a matrix row-wise
`dim(A)` `nrow(A)` `ncol(A)`	Matrix dimension	
`t(A)`	Transpose of a matrix	
`matrix(0,n,m)`	The n × m zero matrix	
`diag (c(s_1,s_2,...,s_n))`	Diagonal matrix with scalars $s_1, s_2,...,s_n$ in the main diagonal. The output is a square matrix of dimension n	If only one positive integer scalar is present (say, s), it produces an identity matrix of dimension s
`diag(A)`	Diagonal extraction of the square matrix **A**	
`sum(diag(A))`	Trace of a square matrix **A**	
`A+B, A−B`	Addition and subtraction of conformable matrices	
`k*A`	Multiplication of a matrix by a scalar k	`A*B` produces an element-wise multiplication of two **conformable** matrices. This *is not* the standard matrix multiplication
`A%*%B`	Matrix multiplication	

Function(s)	Description	Some useful options
`det(A)`	Determinant of a square matrix	
`solve(A)`	Inverse of a square matrix	`solve(A,b)` produces the solution x of the simultaneous system of linear equations `A%*%x=b`
`eigen(A)`	Spectral decomposition of the input matrix. A list with two elements: one vector of eigenvalues, and a matrix of eigenvectors arranged in columns of norm 1 each	The eigenvalues and the eigenvectors are invoked as `eigen(A)` `$values` and `eigen(A)` `$vectors`, respectively
`colMeans(A)` `rowMeans(A)`	Column or row means	Related functions: `colSums(A)`, `rowSums(A)`
`cov(A)`	Given a numeric n × m matrix **A**, creates an m × m variance-covariance matrix	Alternatively, **A** can be data frame
`cor(A)`	Given a numeric n × m matrix **A**, creates an m × m correlation matrix	Alternatively, **A** can be data frame

chapter three

Displaying multivariate data

3.1 The problem of displaying many variables in two dimensions

Graphs must be displayed in two dimensions, either on paper or on a computer screen. It is therefore straightforward to show one variable plotted on a vertical axis against a second variable plotted on a horizontal axis. For example, Figure 3.1 shows the alar extent plotted against the total length for the 49 female sparrows measured by Hermon Bumpus in his study of natural selection, which has been described in Example 1.1. Such plots allow one or more other characteristics of the objects being studied to be shown as well, and in the case of Bumpus' sparrows, survival and nonsurvival are indicated. These plots are simple and can be produced in Excel or another spreadsheet program, as well as in all standard statistical packages. The plots can also be produced using R code, provided in the Appendix to this chapter, for both this type of plot and the more complicated plots that are described below.

It is considerably more complicated to show one variable plotted against another two, but still possible. Thus, Figure 3.2 shows beak and head lengths plotted against total lengths and alar lengths for the 49 sparrows. Again, different symbols are used for survivors and nonsurvivors. This is called a *three-dimensional (3-D) plot* and can be produced in many standard statistical packages and using R code.

It is not possible to show one variable plotted against another three at the same time in some extension of a 3-D plot. Hence, there is a major problem in showing in a simple way the relationships that exist between the individual objects in a multivariate set of data where those objects are each described by four or more variables. Various solutions to this problem have been proposed and are discussed in this chapter.

3.2 Plotting index variables

One approach to making a graphical summary of the differences between objects that are described by more than four variables involves plotting the objects against the values of two or three index variables. Indeed, a major objective of many multivariate analyses is to produce index variables that can be used for this purpose, a process that is sometimes called

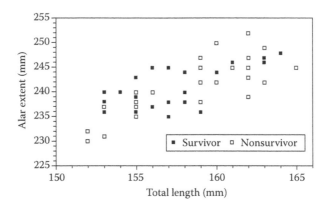

Figure 3.1 Alar extent plotted against total length for the 49 female sparrows measured by Hermon Bumpus.

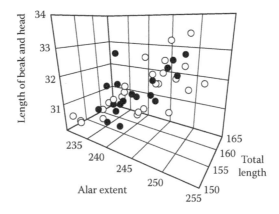

Figure 3.2 Length of the beak and head plotted against the total length and alar extent (all in millimeters) for the 49 female sparrows measured by Hermon Bumpus. ● for survivor, ○ for nonsurvivor.

ordination. For example, principal components, as discussed in Chapter 6, provide one type of index variables. A plot of the values of Principal Component 2 against the values of Principal Component 1 can then be used as a means of representing the relationships between objects graphically, and a display of Principal Component 3 against the first two principal components can also be used if necessary.

The use of suitable index variables has the advantage of reducing the problem of plotting many variables to two or three dimensions, but has the potential disadvantage that some key difference between the objects may be lost in the reduction. This approach is discussed in various different contexts in the chapters that follow and will not be considered further here.

3.3   The draftsman's plot

A draftsman's plot, also called a *scatter plot matrix*, of multivariate data consists of plots of the values for each variable against the values for each of the other variables, with the individual graphs being small enough that they can all be viewed at the same time. This has the advantage of only needing two-dimensional plots, but has the disadvantage that some aspect of the data that is only apparent when three or more variables are considered together will not be apparent.

An example is shown in Figure 3.3. Here, the five variables measured by Hermon Bumpus on 49 sparrows (total length, alar extent, length of beak and head, length of humerus, and length of the keel of the sternum,

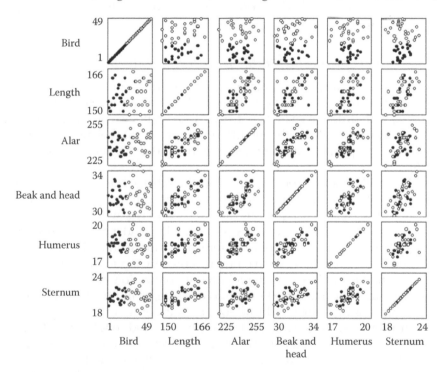

Figure 3.3 Draftsman's plot (scatter plot matrix) of the bird number and five variables measured on 49 female sparrows. The variables are the total length, the alar extent, the length of the beak and head, the length of the humerus, and the length of the keel of the sternum, with obvious abbreviations, all in millimeters. Closed circles represent survivors and open circles nonsurvivors. Only the extreme values are shown on each scale.

all in millimeters) are plotted for the data given in Table 1.1, with an additional first variable being the number of the sparrow, from 1 to 49. Different symbols are used for the measurements on survivors (birds 1–21) and nonsurvivors (birds 22–49). Regression lines are also sometimes added to the plots.

This type of plot is obviously good for showing the relationships between pairs of variables and highlighting the existence of any objects that have unusual values for one or two variables. It can therefore be recommended as part of many multivariate analyses and is available in many statistical packages and using R code. Some packages and R also allow the option of specifying the horizontal and vertical variables without insisting that these are the same.

The individual objects are not easily identified on a draftsman's plot, and it is therefore usually not immediately clear which objects are similar and which are different. Therefore, this type of plot is not suitable for showing relationships between objects, as distinct from relationships between variables.

3.4   The representation of individual data points

An approach to displaying data that is more truly multivariate involves representing each of the objects for which variables are measured by a symbol, with different characteristics of this symbol varying according to different variables. A number of different types of symbol have been proposed for this purpose, including faces (Chernoff, 1973) and stars (Welsch, 1976).

As an illustration, consider the data in Table 1.4 on mean values of six mandible measurements for seven canine groups, as discussed in Example 1.4. Here, an important question concerns which of the other groups is most similar to the prehistoric Thai dog, and it can be hoped that this will become apparent from a graphical comparison of the groups. To this end, Figure 3.4 shows the data represented by faces and stars.

For the faces, there was the following connection between features and the variables: mandible breadth to eye size, mandible height to nose size, length of first molar to brow size, breadth of first molar to ear size, length from first to third molar to mouth size, and length from first to fourth premolars to the amount of smile. For example, the eyes are largest for the Chinese wolf, with the maximum mandible breadth of 13.5 mm, and smallest for the golden jackal, with the minimum mandible length of 8.1 mm. It is apparent from the plots that prehistoric Thai dogs are most similar to modern Thai dogs, and most different from Chinese wolves.

For the stars, the six variables were assigned to rays in the order (1) mandible breadth; (2) mandible height; (3) length of first molar; (4) breadth of first molar; (5) length from first to third molar; and (6) length from first

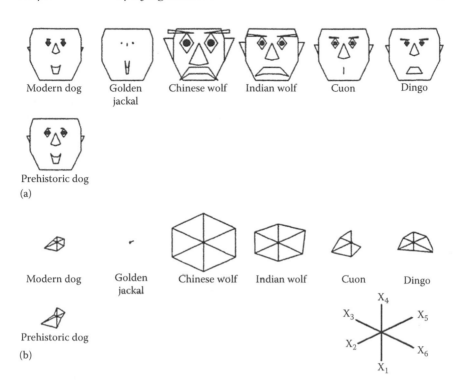

Figure 3.4 Graphical representation of mandible measurements on different canine groups using (a) Chernoff faces and (b) stars.

to fourth premolars. The mandible length is represented by the ray corresponding to six o'clock, and the other variables follow in a clockwise order, as indicated by the key that accompanies the figure. Inspection of the stars indicates again that the prehistoric Thai dogs are most similar to modern Thai dogs and most different from Chinese wolves.

Suggestions for alternatives to faces and stars, and a discussion of the relative merits of different symbols, are provided by Everitt (1978) and Toit et al. (1986, chapter 4). In summary, it can be said that the use of symbols has the advantage of displaying all variables simultaneously, but the disadvantage that the impression gained from the graph may depend quite strongly on the order in which objects are displayed and the order in which variables are assigned to the different aspects of the symbol.

The assignment of variables is likely to have more effect with faces than with stars, because variation in different features of the face may have very different impacts on the observer, whereas this is less likely to be the case with different rays of a star. For this reason, the recommendation is often made that alternative assignments of variables to features should be tried with faces to find what seems to be the best. The subjective nature of this type of process is clearly rather unsatisfactory.

Although the use of faces, stars, and other similar representations for the values of variables on the objects being considered seems to be useful under some circumstances, the fact is that this is seldom done. One reason is the difficulty in finding computer software to produce the graphs. In the past, this software was reasonably easily available, but these options are now very hard to find in statistical packages.

3.5   Profiles of variables

Another way to represent objects described by several variables that are measured on them is by lines that show the profile of variable values. A simple way to draw these involves just plotting the values for the variables, as shown in Figure 3.5 for the seven canine groups that have already been considered. The similarity between prehistoric and modern Thai dogs noted from the earlier graphs is still apparent, as is the difference between prehistoric dogs and Chinese wolves. In this graph, the variables have been plotted in order of their average values for the seven groups to help in emphasizing similarities and differences.

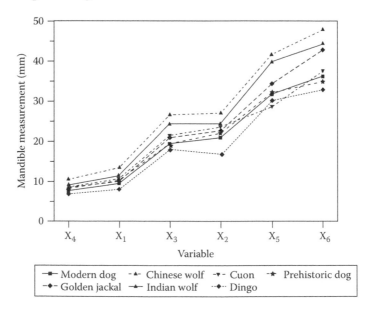

Figure 3.5 Profiles of variables for mandible measurements on seven canine groups. The variables are in order of increasing average values, with X_1 = breadth of mandible, X_2 = height of mandible above the first molar, X_3 = length of first molar, X_4 = breadth of first molar, X_5 = length from first to third molar inclusive, and X_6 = length from first to fourth molar inclusive.

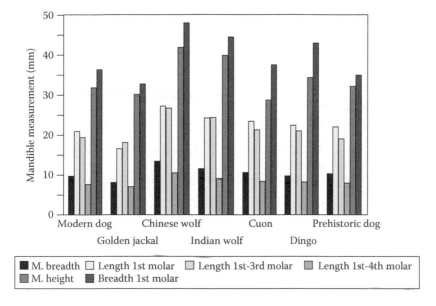

Figure 3.6 An alternative way to show variable profiles, using bars instead of the lines used in Figure 3.5. M = mandible.

An alternative representation, using bars instead of lines, is shown in Figure 3.6. Here, the variables are in their original order because there seems little need to change this when bars are used. The conclusion about similarities and differences between the canine groups is exactly the same as derived from Figure 3.5.

3.6 Discussion and further reading

It seems fair to say that there is no method for displaying data on many variables at a time that is completely satisfactory if it is not desirable to reduce these variables to two or three index variables (using one of the methods to be discussed in Chapters 6 through 12). The three types of method that have been discussed here involve the use of a draftsman's display with all pairs of variables plotted against each other, symbols (stars or faces), and profiles of variables. Which of these is most suitable for a particular application depends on the circumstances, but as a general rule, the draftsman's display is good for highlighting relationships between pairs of variables, while the use of symbols or profiles is good for highlighting unusual cases and similar cases.

For further information about the theory of the construction of graphs in general, see the books by Cleveland (1994) and Tufte (2001). More details on graphical methods specifically for multivariate data are in the books by Everitt (1978), Toit et al. (1986), and Jacoby (1999).

References

Chernoff, H. (1973). Using faces to represent points in K-dimensional space graphically. *Journal of the American Statistical Association* 68: 361–8.

Cleveland, W.S. (1994). *The Elements of Graphing Data*. Revised Edn. Summit, NJ: Hobart.

Everitt, B. (1978). *Graphical Techniques for Multivariate Data*. New York: North-Holland.

Jacoby, W.G. (1999). *Statistical Graphics for Visualizing Multivariate Data*. Thousand Oaks, CA: Sage.

Toit, S.H.C., Steyn, A.G.W. and Stumf, R.H. (1986). *Graphical Exploratory Data Analysis*. New York: Springer.

Tufte, E.R. (2001). *The Visual Display of Quantitative Information*. 2nd Edn. Cheshire, CT: Graphics.

Welsch, R.E. (1976). Graphics for data analysis. *Computers and Graphics* 2: 31–7.

Appendix: Producing Plots in R

A.1   Two-dimensional scatter plots

These are easily produced with the command plot(x,y), where x and y are the variables on the horizontal and the vertical axis, respectively. Points in a scatter plot can be identified according to a factor variable by adding a point character option of the plot function, namely pch=, and conveniently allocating the point symbol to each level of the factor. As an example, the data set *Bumpus sparrows.csv* contains the factor Survivorship, which distinguishes nonsurvivor and survivor sparrows. To implement the identification of points in the scatter plot of the Total_length against the Alar_extent according to survivorship status (as shown in Figure 1.3), indexing is necessary. Remember that factors are stored as integers whose values are determined by the alphabetic order of the factor's labels. Thus, nonsurvivors (identified by the label NS in the factor Survivorship) go first (they are stored as 1) and Survivors (S) get the integer 2. In the R documentation (e.g., for the function points), it is found that the pch code for an open circle is 1 and the code for a filled circle is 16. Thus,

$$\text{plot}\big(\text{Total\_length, Alar\_extent, pch}$$

$$=\text{c}\big(1,16\big)\big[\text{as.numeric}\big(\text{Survivorship}\big)\big]\big)$$

will differentiate nonsurvivors with an open circle (pch = 1) from survivors with a filled circle (pch = 16). For each pair of total length and alar extent, pch is determined by the numeric value of the corresponding level of survivorship.

A.2   Three-dimensional scatter plots

Ligges and Mächler (2003) have developed "scatterplot3d," an R package for the visualization of multivariate data in a 3-D space using parallel projection. The main function is scatterplot3d (the same name as the package). The simplest call to this function is

$$\text{scatterplot3d}\big(\text{x,y,z}\big)$$

where the arguments refer to the variables to be plotted on the x, y, and z axes, respectively. Point identification is also possible, in a similar fashion to that used for a two-dimensional scatter plot using plot(x,y). An example with a slightly different version of Figure 3.2 is given in the supplementary material found in the book's website.

A.3   Draftsman's plots

R has a way to produce a matrix of scatter plots using a basic command called pairs(). As an example, in Figure 3.3, a draftsman's plot for the Bumpus sparrows data, the variable names are shown on the main diagonal, and the points are identified by their Survivorship status. The user may want to display histograms, boxplots, density plots, etc. on the diagonal, but some programming skills are needed. There are several ways to circumvent the programming hassles, all of them involving R packages. Thus, the SciViews package (Grosjean, 2016) offers the inclusion of the argument diag.panel= to the function pairs(), as an easy way to specify the diagonal elements (e.g., diag.panel= "hist" for histograms, diag.panel= "boxplot", etc.). Another friendlier function allowing customized draftsman's plots is scatterplotMatrix, present in the car package (Fox and Weisberg, 2011). More sophisticated functions for similar purposes are given in the packages lattice (Deepayan, 2008) and GGally (Schloerke et al., 2014).

A.4   Representation of individual data points: Chernoff faces and stars

Two R packages that provide functions to make Chernoff faces are aplpack (Wolf and Bielefeld, 2014) and TeachingDemos (Snow, 2013). Through the single function *faces*, the aplpack package allows complex displays (e.g., colored face regions, scatter plots of two variables (X,Y) where faces are placed in the (X,Y) coordinates, etc.). TeachingDemos, in turn, offers two functions: faces, a simplified version of that in aplpack, and faces2, which requires the input data to be provided as a matrix. For more details, read the documentation and examples provided in the help files for these two packages. Examples of the use of faces and faces2 for the exploration of the Canine data are given in the supplementary material of this book.

For Stars, users just need to invoke the function stars() from the core package graphics. In addition to the input matrix or data frame, further arguments can be given to get variants of the basic star plot, such as segment plots (circular sectors) and radar plots. See the R documentation for more details. Using the Canine data, we provide on the book's website an example on the use of the function stars as a way to produce a plot similar to Figure 4.3b, and a variant in which circular sectors are displayed for each variable.

A.5   Profiles of variables

Profile plots can be generated with a combination of a high-level graphics function (plot()) and low-level graphics functions (e.g., axis() and `lines()`. Low-level functions in R cannot be invoked if a high-level function has not been called earlier. This strategy assures that additional elements are placed in the plot without modifying the basic graph layout determined by the high-level function. As variables are plotted in order of their average values, R functions `order()` and `rank()` are also needed. R scripts using all these commands can be found in the book's website, producing plots similar to those shown in Figures 3.5 and 3.6, but with the latter plot using bars (via the `barplot()` function) instead of lines.

References

Deepayan, S. (2008). *Lattice: Multivariate Data Visualization with R.* New York: Springer.

Fox, J. and Weisberg, S. (2011). *An R Companion to Applied Regression.* 2nd Edn. Thousand Oaks, CA: Sage.

Grosjean, Ph. (2016). *SciViews: A GUI API for R.* Mons, Belgium: UMONS. http://www.sciviews.org/SciViews-R

Ligges, U. and Mächler, M. (2003). Scatterplot3d: An R package for visualizing multivariate data. *Journal of Statistical Software* 8(11): 1–20.

Schloerke, B., Crowley, J., Cook, D., Hofmann, H., Wickham, H., Briatte, F., Marbach, M. and Thoen, E. (2014). GGally: Extension to ggplot2. R package version 0.5.0. http://CRAN.R-project.org/package=GGally

Snow, G. (2013). TeachingDemos: Demonstrations for teaching and learning. R package version 2.9. http://CRAN.R-project.org/package=TeachingDemos

Wolf, H.P. and Bielefeld, U. (2014). Aplpack: Another plot PACKage: Stem.leaf, bagplot, faces, spin3R, plotsummary, plothulls, and some slider functions. R package version 1.3.0. http://CRAN.R-project.org/package=aplpack

chapter four

Tests of significance with multivariate data

4.1 Simultaneous tests on several variables

When data are collected for several variables on the same sample units, then it is always possible to examine the variables one at a time as far as tests of significance are concerned. For example, if the sample units are in two groups, then a difference between the means for the two groups can be tested separately for each variable. Unfortunately, there is a drawback to this simple approach because of the repeated use of significance tests, each of which has a certain probability of leading to a wrong conclusion. As will be discussed further in Section 4.4, the probability of falsely finding at least one significant difference accumulates with the number of tests carried out, so that it may become unacceptably large.

There are ways of adjusting significance levels to allow for many tests being carried out at the same time, but it may be preferable to conduct a single test that uses the information from all variables together. For example, it might be desirable to test the hypothesis that the means of all variables are the same for two multivariate populations, with a significant result being taken as evidence that the means differ for at least one variable. These types of overall test are considered in this chapter for the comparison of means and the comparison of variation for two or more samples.

4.2 Comparison of mean values for two samples: The single-variable case

Consider the data in Table 1.1 on the body measurements of 49 female sparrows. Consider in particular the first measurement, which is total length. A question of some interest might be whether the mean of this variable was the same for survivors and nonsurvivors of the storm that led to the birds being collected. There is, then, a sample of 21 survivors and a second sample of 28 nonsurvivors. Assuming that these are effectively random samples from much larger populations of survivors and nonsurvivors, the question then is whether the two sample means are significantly different in the sense that the observed mean difference is so large that it is unlikely to have occurred by chance if the population means are equal. A standard approach would be to carry out a t-test.

Thus, suppose that in a general situation, there is a single variable X, and two random samples of values are available from different populations. Let x_{i1} denote the values of X in the first sample, for $i = 1, 2, ..., n_1$, and x_{i2} denote the values in the second sample, for $i = 1, 2, ..., n_2$. Then, the mean and variance for the jth sample are

$$\bar{x}_j = \sum_{i=1}^{n_j} x_{ij}/n$$

and

$$s_j^2 = \sum_{i=1}^{n_j} \left(x_{ij} - \bar{x}_j \right)^2 / \left(n_j - 1 \right) \tag{4.1}$$

On the assumption that X is normally distributed in both samples, with a common within-sample variance, a test to see whether the two sample means are significantly different involves calculating the statistic

$$t = \left(\bar{x}_1 - \bar{x}_2 \right) / \left\{ s \cdot \sqrt{(1/n_1 + 1/n_2)} \right\} \tag{4.2}$$

and seeing whether this is significantly different from zero in comparison with the t distribution with $n_1 + n_2 - 2$ degrees of freedom (df). Here,

$$s^2 = \left\{ (n_1 - 1)s_1^2 + (n_2 - 1)s_2^2 \right\} / (n_1 + n_2 - 2) \tag{4.3}$$

is the pooled estimate of variance from the two samples.

It is known that this test is fairly robust to the assumption of normality, so that, provided that the population distributions of X are not too different from normal, it should be satisfactory, particularly for sample sizes of about 20 or more. The assumption of equal population variances is also not too crucial if the ratio of the true variances is within the limits 0.4–2.5. The test is particularly robust if the two sample sizes are equal, or nearly so.

If there are no concerns about nonnormality but the population variances may be unequal, then one possibility is to use a modified t-test. For example, Welch's (1951) test can be used: this has the test statistic

$$t = \left(\bar{x}_1 - \bar{x}_2 \right) / \left\{ \sqrt{\left(s_1^2/n_1 + s_2^2/n_2 \right)} \right\} \tag{4.4}$$

Evidence for unequal population means is then obtained if t is significantly different from zero in comparison with the t distribution with the
df equal to

$$v = (w_1 + w_2)^2 / \{w_1^2/(n_1 - 1) + w_2^2/(n_2 - 1)\} \qquad (4.5)$$

where:
$$w_1 = s_1^2/n_1$$
$$w_2 = s_2^2/n_2$$

When there are concerns about both nonnormality and unequal
variances, it has been shown by Manly and Francis (2002) that it may not
be possible to reliably test for a difference in the population means. In
particular, there may be too many significant results, providing evidence
for a population mean difference when this does not really exist, irrespective of what test procedure is used. Manly and Francis provided a solution for this problem via a testing scheme that includes an assessment of
whether two or more samples differ with respect to means or variances
using randomization tests (Manly, 2009, section 4.6), and also an assessment of whether the randomization test for mean differences is reliable.
See their paper for more details.

4.3   Comparison of mean values for two samples: The multivariate case

Consider again the female sparrow data that are shown in Table 1.1. The
t-test described in the previous section can obviously be employed for
each of the five measurements shown in the table (total length, alar extent,
length of beak and head, length of humerus, and length of keel of sternum). In that way, it is possible to decide which, if any, of these variables
appear to have had different mean values for the populations of survivors
and nonsurvivors. However, in addition to these tests, it may also be of
some interest to know whether all five variables considered together suggest a difference between survivors and nonsurvivors. In other words,
does the total evidence point to mean differences between the populations of surviving and nonsurviving sparrows?

What is needed to answer this question is a multivariate test. One possibility is Hotelling's T^2-test. The statistic used is then a generalization of the
t-statistic of Equation 4.2 or, to be more precise, the square of this t-statistic.

In a general case, there will be p variables $X_1, X_2, \ldots, X_p$ being considered, and two samples with sizes n_1 and n_2. There are then two sample
mean vectors, $\bar{x}_1$ and $\bar{x}_2$, and two sample covariance matrices, C_1 and C_2,
with these being calculated as explained in Section 2.7.

Assuming that the population covariance matrices are the same for both populations, a pooled estimate of this matrix is

$$C = \{(n_1 - 1)C_1 + (n_2 - 1)C_2\}/(n_1 + n_2 - 2) \tag{4.6}$$

Hotelling's T^2-statistic is then defined as

$$T^2 = n_1 \cdot n_2 (\bar{x}_1 - \bar{x}_2)' C^{-1} (\bar{x}_1 - \bar{x}_2)/(n_1 + n_2) \tag{4.7}$$

A significantly large value for this statistic is evidence that the two population mean vectors are different. The significance, or lack of significance, of T^2 is most simply determined by using the fact that if the null hypothesis of equal population mean vectors is true, then the transformed statistic

$$F = (n_1 + n_2 - p - 1) \ T^2 / \{(n_1 + n_2 - 2)p\} \tag{4.8}$$

follows an F distribution with p and $(n_1 + n_2 - p - 1)$ df.

The T^2-statistic is a quadratic form, as defined in Section 2.5. It can therefore be written as the double sum

$$T^2 = \{(n_1 n_2)/(n_1 + n_2)\} \sum_{i=1}^{p} \sum_{k=1}^{p} (\bar{X}_{1i} - \bar{X}_{2i}) c^{ik} (\bar{X}_{1k} - \bar{X}_{2k}) \tag{4.9}$$

which may be simpler to compute. Here, $\bar{x}_{ji}$ is the mean of variable X_i in the jth sample, and c^{ik} is the element in the ith row and kth column of the inverse matrix C^{-1}.

The two samples being compared using the T^2-statistic are assumed to come from multivariate normal distributions with equal covariance matrices. Some deviation from multivariate normality is probably not serious. A moderate difference between population covariance matrices is also not too important, particularly with equal or nearly equal sample sizes. If the two population covariance matrices are very different, and sample sizes are very different as well, then a modified test can be used (Yao, 1965), but this still relies on the assumption of multivariate normality.

Example 4.1: Testing mean values for Bumpus' female sparrows

As an example of the use of the univariate and multivariate tests that have been described for two samples, consider again the female sparrow data from Table 1.1. Here, there is a question about whether there are any differences between survivors and nonsurvivors with respect to the mean values of five morphological characters.

Table 4.1 Comparison of mean values for survivors and nonsurvivors for Bumpus' female sparrows with variables taken one at a time

Variable	Survivors		Nonsurvivors		t (47 df)	p-value[a]
	x_1	s_1^2	x_2	s_2^2		
Total length	157.38	11.05	158.43	15.07	−0.99	0.327
Alar extent	241.00	17.50	241.57	32.55	−0.39	0.698
Length of beak and head	31.43	0.53	31.48	0.73	−0.20	0.842
Length of humerus	18.50	0.18	18.45	0.43	0.33	0.743
Length of keel of sternum	20.81	0.58	20.84	1.32	−0.10	0.921

[a] Probability of obtaining a t-value as far from zero as the observed value if the null hypothesis of no population mean difference is true.

First of all, tests on the individual variables can be considered, starting with X_1, the total length. The mean of this variable for the 21 survivors is $\bar{x}_1 = 157.38$, while the mean for the 28 nonsurvivors is $\bar{x}_2 = 158.43$. The corresponding sample variances are $s_1^2 = 11.05$ and $s_2^2 = 15.07$. The pooled variance from Equation 4.3 is therefore

$$s^2 = (20 \times 11.05 + 27 \times 15.07)/47 = 13.36$$

and the t-statistic of Equation 4.2 is

$$t = (157.38 - 158.43) \Big/ \sqrt{\{13.36(1/21 + 1/28)\}} = -0.99$$

with $n_1 + n_2 - 2 = 47$ df. This is not significantly different from zero at the 5% level, so there is no evidence of a population mean difference between survivors and nonsurvivors with regard to total length.

Table 4.1 summarizes the results of tests on all five of the variables taken individually. In no case is there any evidence of a population mean difference between survivors and nonsurvivors.

For tests on all five variables considered together, it is necessary to know the sample mean vectors and covariance matrices. The means are given in Table 4.1, and the covariance matrices are defined in Section 2.7. For the sample of 21 survivors, the mean vector and covariance matrix are

$$\bar{x}_1 = \begin{bmatrix} 157.381 \\ 241.000 \\ 31.433 \\ 18.500 \\ 20.810 \end{bmatrix} \text{ and } C_1 = \begin{bmatrix} 11.048 & 9.100 & 1.557 & 0.870 & 1.286 \\ 9.100 & 17.500 & 1.910 & 1.310 & 0.880 \\ 1.557 & 1.910 & 0.531 & 0.189 & 0.240 \\ 0.870 & 1.310 & 0.189 & 0.176 & 0.133 \\ 1.286 & 0.880 & 0.240 & 0.133 & 0.575 \end{bmatrix}$$

For the sample of 28 nonsurvivors, the results are

$$\bar{x}_2 = \begin{bmatrix} 158.429 \\ 241.571 \\ 31.479 \\ 18.446 \\ 20.839 \end{bmatrix} \text{ and } C_2 = \begin{bmatrix} 15.069 & 17.190 & 2.243 & 1.746 & 2.931 \\ 17.190 & 32.550 & 3.398 & 2.950 & 4.066 \\ 2.243 & 3.398 & 0.728 & 0.470 & 0.559 \\ 1.746 & 2.950 & 0.470 & 0.434 & 0.506 \\ 2.931 & 4.066 & 0.559 & 0.506 & 1.321 \end{bmatrix}$$

The pooled sample covariance matrix is then

$$C = (20 \cdot C_1 + 27 \cdot C_2)/47 = \begin{bmatrix} 13.358 & 13.748 & 1.951 & 1.373 & 2.231 \\ 13.748 & 26.146 & 2.765 & 2.252 & 2.710 \\ 1.951 & 2.765 & 0.645 & 0.350 & 0.423 \\ 1.373 & 2.252 & 0.350 & 0.324 & 0.347 \\ 2.231 & 2.710 & 0.423 & 0.347 & 1.004 \end{bmatrix}$$

where, for example, the element in the second row and third column is

$$(20 \times 1.910 + 27 \times 3.398)/47 = 2.765$$

The inverse of the matrix C is found to be

$$C^{-1} = \begin{bmatrix} 0.2061 & -0.0694 & -0.2395 & 0.0785 & -0.1969 \\ -0.0694 & 0.1234 & -0.0376 & -0.5517 & 0.0277 \\ -0.2395 & -0.0376 & 4.2219 & -3.2624 & -0.0181 \\ 0.0785 & -0.5517 & -3.2624 & 11.4610 & -1.2720 \\ -0.1969 & 0.0277 & -0.0181 & -1.2720 & 1.8068 \end{bmatrix}$$

This can be verified by evaluating the product $C \times C^{-1}$ and seeing that this is a unit matrix, apart from rounding errors.

Substituting the elements of C^{-1} and other values into Equation 4.7 produces

$$T^2 = \{(21\times28)(21+28)\} \left[(157.381-158.429)\times0.2061\times(157.381 \right.$$
$$-158.429-(157.318-158.429)\times0.0694\times(241.000-241.571)$$
$$\left. +(20.810-20.839)\times1.8068\times(20.810-20.839)\right] = 2.824.$$

Using Equation 4.8, this converts to an F-statistic of

$$F = (21+28-5-1)\times\ 2.824/\{(21+28\ -2)\times5\} = 0.517$$

with 5 and 43 df. Clearly, this is not significantly large, because a significant F-value must exceed unity. Hence, there is no evidence of a difference in population means for survivors and non-survivors, taking all five variables together.

4.4   Multivariate versus univariate tests

In the last example, there were no significant results either for the variables considered individually or for the overall multivariate test. However, it is quite possible to have insignificant univariate tests but a significant multivariate test. This can occur because of the accumulation of the evidence from the individual variables in the overall test. Conversely, an insignificant multivariate test can occur when some univariate tests are significant, because the evidence of a difference provided by the significant variables is swamped by the evidence of no difference provided by the other variables.

One important aspect of the use of a multivariate test as distinct from a series of univariate tests concerns the control of type one error rates. A type one error involves finding a significant result when, in reality, the two samples being compared come from populations with the same mean (for a univariate test) or means (for a multivariate test). With a univariate test at the 5% level, there is a 0.95 probability of a nonsignificant result when the population means are the same. Hence, if p independent tests are carried out under these conditions, then the probability of getting no significant results is 0.95^p. The probability of at least one significant result is therefore $1 - 0.95^p$, which may be unacceptably large. For example, if $p=5$, then the probability of at least one significant result by chance alone is $1-0.95^5=0.23$. With multivariate data, variables are usually not independent, so $1 - 0.95^p$ does not quite give the correct probability of at least one significant result by chance alone if variables are tested one by one with univariate t-tests. However, the principle still applies that the more tests that are made, the higher is the probability of obtaining at least one significant result by chance.

On the other hand, a multivariate test such as Hotelling's T^2 test using the 5% level of significance gives a 0.05 probability of a type one error, irrespective of the number of variables involved, providing that the assumptions of the test hold. This is a distinct advantage over a series of univariate tests, particularly when the number of variables is large.

There are ways of adjusting significance levels to control the overall probability of a type one error when several univariate tests are carried out. The simplest approach involves using a Bonferroni adjustment. For example, if p univariate tests are carried out using the significance level $(5/p)\%$, then the probability of obtaining any significant results is less than 0.05 if the null hypothesis is true for each test. More generally, if p tests are carried out using the significance level $(100\alpha/p)\%$, then the probability of obtaining any significant results by chance is less than α.

Some people are not inclined to use a Bonferroni correction to significance levels, because the significance levels applied to the individual tests become so extreme if p is large. For example, with $p = 10$ and an overall 5% level of significance, a univariate test result is only declared significant if it is significant at the 0.5% level. This has led to the development of some slightly less conservative variations on the Bonferroni correction, as discussed by Manly (2009, section 4.9).

It can certainly be argued that the use of a single multivariate test provides a better procedure in many cases than making a large number of univariate tests. A multivariate test also has the added advantage of taking proper account of the correlation between variables.

4.5   Comparison of variation for two samples: The single-variable case

With a single variable, the best-known method for comparing the variation in two samples is the F-test. If s_j^2 is the variance in the jth sample, calculated as shown in Equation 4.1, then the ratio s_1^2/s_2^2 is compared with percentage points of the F distribution with (n_1-1) and (n_2-1) df. A value of the ratio that is significantly different from one is then evidence that the samples are from two populations with different variances. Unfortunately, the F-test is known to be rather sensitive to the assumption of normality. A significant result may well be due to the fact that a variable is not normally distributed rather than to unequal variances. For this reason, it is sometimes argued that the F-test should never be used to compare variances.

A robust alternative to the F-test is Levene's (1960) test. The idea here is to transform the original data in each sample into absolute deviations from the sample mean or the sample median and then test for a significant difference between the mean deviations in the two samples, using a t-test. Although absolute deviations from the sample means are sometimes used, a more robust test is likely to be obtained by using absolute deviations from the sample medians (Schultz, 1983). The procedure using medians is illustrated in Example 4.2.

4.6 Comparison of variation for two samples: The multivariate case

Many computer packages use Box's M-test to compare the variation in two or more multivariate samples. Because this applies for two or more samples, it is described in Section 4.8. This test is known to be rather sensitive to the assumption that the samples are from multivariate normal distributions. There is, therefore, always the possibility that a significant result is due to nonnormality rather than to unequal population covariance matrices.

An alternative procedure that should be more robust can be constructed using the principle behind Levene's test. Thus, the data values can be transformed into absolute deviations from sample means or medians. The question of whether the two samples display significantly different amounts of variation is then transformed into a question of whether the transformed values show significantly different mean vectors. Testing of the mean vectors can be done using a T^2-test.

Another possibility was suggested by Van Valen (1978). This involves calculating

$$d_{ij} = \sqrt{\left\{ \sum_{k=1}^{p} \left(x_{ijk} - \bar{x}_{jk} \right)^2 \right\}} \tag{4.10}$$

where:

x_{ijk} is the value of variable X_k for the ith individual in sample j

$\bar{x}_{jk}$ is the mean of the same variable in the sample

The sample means of the d_{ij} values are compared with a t-test. Obviously, if one sample is more variable than another, then the mean d_{ij} values will tend to be higher in the more variable sample.

To ensure that all variables are given equal weight, each variable should be standardized so that the mean is zero and variance is one for all samples combined before the calculation of the d_{ij} values. For a more robust test, it may be better to use sample medians in place of the sample means in Equation 4.10. Then, the formula for d_{ij} values is

$$d_{ij} = \sqrt{\left\{ \sum_{k=1}^{p} \left(x_{ijk} - M_{jk} \right)^2 \right\}} \tag{4.11}$$

where M_{jk} is the median for variable X_k in the jth sample.

The T^2-test and Van Valen's test for deviations from medians are illustrated in the example that follows. One point to note about the use of the test statistics 4.10 and 4.11 is that they are based on an implicit assumption that if the two samples being tested differ, then one sample will be more variable than the other for all variables. A significant result cannot be expected in a case where, for example, X_1 and X_2 are more variable in sample 1, but X_3 and X_4 are more variable in sample 2. The effect of the differing variances would then tend to cancel out in the calculation of d_{ij}. Thus, Van Valen's test is not appropriate for situations where changes in the level of variation are not expected to be consistent for all variables.

Example 4.2: Testing variation for female sparrows

With Bumpus' female sparrow data shown in Table 1.1, one interesting question concerns whether the nonsurvivors are more variable than the survivors. This is what would be expected if stabilizing selection took place.

To examine this question, first of all, the individual variables can be considered one at a time, starting with X_1, the total length. For Levene's test, the original data values are transformed into deviations from sample medians. The median for survivors is 157 mm, and the absolute deviations from this median for the 21 birds in the sample then have a mean of $\bar{x}_1 = 2.571$ and a variance of $s_1^2 = 4.257$. The median for nonsurvivors is 159 mm, and the absolute deviations from this median for the 28 birds in the sample have a mean of $\bar{x}_2 = 3.286$ with a variance of $s_2^2 = 4.212$. The pooled variance from Equation 4.3 is 4.231, and the t-statistic of Equation 4.2 is

$$t = (2.57 - 3.29)/\left\{4.231(1/21 + 1/28)\right\}^{\frac{1}{2}} = -1.20$$

with 47 df.

Because nonsurvivors would be more variable than survivors if stabilizing selection occurred, it is a one-sided test that is required here, with low values of t providing evidence of selection. The observed value of t is not significantly low in the present instance. The t-values for the other variables are as follows: the alar extent, $t = -1.18$; the length of the beak and head, $t = -0.81$; the length of the humerus, $t = -1.91$; and the length of keel of the sternum, $t = -1.41$. Only for the length of the humerus is the result significantly low at the 5% level.

Table 4.2 shows the absolute deviations from sample medians for the data after they have been standardized for Van Valen's test. For example, the first value given for variable 1, for survivors, is 0.274. This was obtained as follows. First, the original data were coded to have a zero mean and a unit variance for all 49 birds. This transformed the total length for the first survivor to $(156 - 157.980)/3.654 = -0.542$. The median transformed length for survivors was then -0.268. Hence, the absolute deviation from the sample median for the first survivor is $|-0.542 - (-0.268)| = 0.274$, as shown in Table 4.2.

Table 4.2 Standardized values for the data on the size of the surviving (1) and nonsurviving (2) female sparrows as shown in Table 1.1

	Standardized data					Absolute deviations from sample medians					
Survival	Total length	Alar extent	Beak and head	Length of humerus	Keel of sternum	Total length	Alar extent	Beak and head	Length of humerus	Keel of sternum	d
1	-0.542	0.725	0.177	0.054	-0.329	0.274	0.987	0.252	0.000	0.101	1.059
1	-1.089	-0.262	-1.333	-1.009	-1.237	0.821	0.000	1.258	1.063	1.009	2.099
1	-1.363	-0.262	-0.578	-0.123	-0.229	1.095	0.000	0.503	0.177	0.000	1.218
1	-1.363	-1.051	-0.704	-1.363	-0.632	1.095	0.789	0.629	1.418	0.403	2.095
1	-0.815	0.330	0.051	0.231	-0.531	0.547	0.592	0.126	0.177	0.303	0.888
1	1.374	1.120	0.680	0.940	0.074	1.642	1.381	0.755	0.886	0.303	2.460
1	-0.268	-0.656	-0.704	-0.123	-0.632	0.000	0.395	0.629	0.177	0.403	0.864
1	-0.815	-0.459	1.687	0.231	0.377	0.547	0.197	1.762	0.177	0.605	1.959
1	1.647	1.317	1.561	1.118	0.276	1.916	1.579	1.636	1.063	0.504	3.197
1	0.006	-0.656	-0.578	0.586	1.184	0.274	0.395	0.503	0.532	1.412	1.662
1	0.006	-0.262	-0.200	0.231	1.184	0.274	0.000	0.126	0.177	1.412	1.455
1	0.553	0.528	-0.452	0.231	-0.329	0.821	0.789	0.377	0.177	0.101	1.217
1	0.827	0.922	1.058	1.472	0.982	1.095	1.184	1.132	1.418	1.210	2.712
1	-0.268	0.725	0.680	1.118	-0.834	0.000	0.987	0.755	1.063	0.605	1.744
1	-0.268	-1.248	0.051	-0.655	-1.035	0.000	0.987	0.126	0.709	0.807	1.464
1	-0.542	-0.854	-0.704	-0.832	-0.531	0.274	0.592	0.629	0.886	0.303	1.303
1	0.006	0.528	-0.074	0.054	0.780	0.274	0.789	0.000	0.000	1.009	1.310
1	-1.363	-0.656	-1.207	-0.477	0.074	1.095	0.395	1.132	0.532	0.303	1.735
1	-0.815	-1.051	-1.459	0.054	-0.733	0.547	0.789	1.384	0.000	0.504	1.759
1	1.374	0.922	1.310	0.231	1.083	1.642	1.184	1.384	0.177	1.311	2.786
1	0.279	-1.051	0.051	-0.832	0.679	0.547	0.789	0.126	0.886	0.908	1.596

(*Continued*)

Table 4.2 (Continued) Standardized values for the data on the size of the surviving (1) and nonsurviving (2) female sparrows as shown in Table 1.1

Survival	Standardized data					Absolute deviations from sample medians					
	Total length	Alar extent	Beak and head	Length of humerus	Keel of sternum	Total length	Alar extent	Beak and head	Length of humerus	Keel of sternum	d
										Mean	1.742
										Var	0.402
Median	−0.268	−0.262	−0.074	0.054	−0.229						
2	−0.815	−0.262	−0.074	−0.832	−0.128	1.095	0.395	0.126	0.886	0.000	1.468
2	−0.542	−0.262	0.051	−0.477	−0.229	0.821	0.395	0.000	0.532	0.101	1.059
2	0.553	0.133	1.435	0.586	0.881	0.274	0.000	1.384	0.532	1.009	1.814
2	−1.636	−1.840	−1.459	−2.250	−1.035	1.916	1.973	1.510	2.304	0.908	3.997
2	0.553	1.711	0.303	0.586	1.688	0.274	1.579	0.252	0.532	1.816	2.492
2	−0.815	−0.854	−0.578	0.054	−0.834	1.095	0.987	0.629	0.000	0.706	1.751
2	−0.268	0.725	0.932	1.826	0.578	0.547	0.592	0.881	1.772	0.706	2.251
2	1.921	0.725	2.065	2.358	1.890	1.642	0.592	2.013	2.304	2.017	4.059
2	−1.363	−2.038	−1.710	−2.072	−1.035	1.642	2.171	1.762	2.127	0.908	3.982
2	1.100	−0.459	−1.459	−0.832	2.293	0.821	0.592	1.510	0.886	2.421	3.154
2	1.100	0.330	0.177	0.586	0.478	0.821	0.197	0.126	0.532	0.605	1.174

2	0.279	0.725	0.429	0.054	0.881	0.000	0.592	0.377	0.000	1.009	1.229
2	0.279	1.120	-0.704	-0.655	-1.842	0.000	0.987	0.755	0.709	1.715	2.233
2	-0.815	0.330	-0.704	0.054	0.478	1.095	0.197	0.755	0.000	0.605	1.474
2	1.100	2.106	0.555	1.118	1.385	0.821	1.973	0.503	1.063	1.513	2.871
2	-1.636	-2.235	-1.333	-2.072	-2.246	1.916	2.368	1.384	2.127	2.118	4.495
2	0.279	0.133	-0.829	-0.477	-0.329	0.000	0.000	0.881	0.532	0.202	1.048
2	-0.815	-0.656	-0.326	-1.009	-1.540	1.095	0.789	0.377	1.063	1.412	2.256
2	1.374	1.514	2.442	1.826	1.991	1.095	1.381	2.391	1.772	2.118	4.056
2	1.374	0.133	-0.578	-0.655	-0.128	1.095	0.000	0.629	0.709	0.000	1.448
2	-0.542	-0.854	0.303	-0.477	-0.531	0.821	0.987	0.252	0.532	0.403	1.468
2	0.279	-0.656	0.051	-0.123	-0.531	0.000	0.789	0.000	0.177	0.403	0.904
2	0.827	0.725	0.806	1.118	-0.027	0.547	0.592	0.755	1.063	0.101	1.536
2	-0.815	-1.248	-0.955	-1.363	-1.237	1.095	1.381	1.007	1.418	1.110	2.713
2	1.100	1.120	0.555	1.118	-0.430	0.821	0.987	0.503	1.063	0.303	1.767
2	-1.363	-0.854	-1.081	0.231	-0.430	1.642	0.987	1.132	0.177	0.303	2.253
2	1.100	0.725	1.310	0.054	0.276	0.821	0.592	1.258	0.000	0.403	1.664
2	1.647	1.317	1.058	0.586	0.074	1.368	1.184	1.007	0.532	0.202	2.147
Median	0.279	0.133	0.051	0.054	-0.128					Mean	2.242
										Var	1.110

Note: The standardized values have a mean of zero and a standard deviation of one for all of the data combined, for all five variables. These values are used for the Van Valen (1978) test to compare the variation in the two samples of survivors and nonsurvivors based on deviations from sample medians and Equation 4.11. The Van Valen test compares the two sample means of d-values using a t-test. This shows that there is significantly more variation in size for nonsurvivors than for survivors ($t = -1.92$ with 47 df, $p = 0.030$ on a one-sided test).

Comparing the transformed sample mean vectors for the five variables using Hotelling's T^2-test gives a test statistic of $T^2 = 2.82$, corresponding to an F-statistic of 0.52 with 5 and 43 df using Equation 4.8. There is, therefore, no evidence of a significant difference between the samples from this test, because the F-value is less than one, with a significance level of 0.76.

Finally, consider Van Valen's test. The d-values from Equation 4.11, that is, the sums on the right-hand side of the equation for individuals within samples, are shown in the last column of Table 4.2. The mean for survivors is 1.742, with variance 0.402. The mean for nonsurvivors is 2.242, with variance 1.110. The t-value from Equation 4.2 is then −1.92, which is significantly low at the 5% level (p = 0.03 on a one-sided test). Hence, this test indicates more variation for nonsurvivors than for survivors.

An explanation for the significant result with this test, but no significant result with Levene's test, is not hard to find. As noted above, Levene's test is not directional, and does not take into account the expectation that the survivors will, if anything, be less variable than the nonsurvivors. On the other hand, Van Valen's test is specifically for less variation in Sample 1 than in Sample 2, for all variables. In the present case, all the variables show less variation in Sample 1 than in Sample 2. Van Valen's test has emphasized this fact, but Levene's test has not.

4.7   Comparison of means for several samples

When there is a single variable and several samples to be compared, the generalization of the t-test is the F-test from a one-factor analysis of variance. The calculations are as shown in Table 4.3.

When there are several variables and several samples, the situation is complicated by the fact that there are four alternative statistics that are

Table 4.3 One-factor analysis of variance for comparing the mean values of samples from m populations, with a single variable

Source of variation	Sum of squares	df	Mean square	F-ratio
Between samples	$B = T - W$	$m - 1$	$M_B = B/ (m - 1)$	$F = M_B/ M_W$
Within samples	$W = \sum\limits_{j=1}^{m} \sum\limits_{i=1}^{n_j} \left(x_{ij} - x_j\right)^2$	$n = m$	$M_W = W/ (n - m)$	
Total	$T = \sum\limits_{j=1}^{m} \sum\limits_{i=1}^{n_j} \left(x_{ij} - x\right)^2$	$n = 1$		

Note: n_j = size of the jth sample; $n = n_1 + n_2 + \ldots + n_m$ = total number of observations; x_{ij} = ith observation in the jth sample; x_j = mean of the jth sample; x = mean of all observations.

commonly used to test the hypothesis that all the samples came from populations with the same mean vector.

The first test to be considered uses Wilks' lambda statistic

$$\Lambda = |\mathbf{W}| / |\mathbf{T}| \tag{4.12}$$

where:
  $|\mathbf{W}|$ is the determinant of the within-sample sum of squares and cross products matrix
  $|\mathbf{T}|$ is the determinant of the total sum of squares and cross products matrix

Essentially, this compares the variation within the samples with the variation both within and between the samples. Here, the matrices $\mathbf{T}$ and $\mathbf{W}$ require some further explanation. Let x_{ijk} denote the value of variable X_k for the ith individual in the jth sample, let $\bar{x}_{jk}$ denote the mean of X_k in the same sample, and let $\bar{x}_k$ denote the overall mean of X_k for all the data taken together. In addition, assume that there are m samples, with the jth of size n_j. Then, the element in row r and column c of $\mathbf{T}$ is

$$t_{rc} = \sum_{j=1}^{m} \sum_{i=1}^{n_j} \left(x_{ijr} - \bar{x}_r \right)\left(x_{ijc} - \bar{x}_c \right) \tag{4.13}$$

and the element in row r and column c of $\mathbf{W}$ is

$$w_{rc} = \sum_{j=1}^{m} \sum_{i=1}^{n_j} \left(x_{ijr} - \bar{x}_{jr} \right)\left(x_{ijc} - \bar{x}_{jc} \right) \tag{4.14}$$

What is meant by a determinant is briefly discussed in Section 2.4. Here, all that needs to be known is that they are scalar quantities, that is, ordinary numbers rather than vectors or matrices, and that special computer algorithms are needed to calculate them unless the matrices involved are of size 2×2, or possibly 3×3.

If Λ is small, this indicates that the variation within the samples is low in comparison with the total variation. This, then, provides evidence that the samples do not come from populations with the same mean vector. An approximate test for whether the within-sample variation is significantly low in this respect is described in Table 4.4. Tables of exact critical values are also available.

Let $\lambda_1 \geq \lambda_2 \geq \ldots \geq \lambda_p \geq 0$ be the eigenvalues of $\mathbf{W}^{-1}\mathbf{B}$, where $\mathbf{B} = \mathbf{T} - \mathbf{W}$ is called the between-sample matrix of sums of squares and cross products, because the typical entry is the difference between a total sum of squares or cross product minus the corresponding term within samples. Then, Wilks' lambda can also be expressed as

$$\Lambda = \prod_{i=1}^{p} 1/(1+\lambda_i) \tag{4.15}$$

which is the form that is sometimes used to represent it.

A second statistic is the largest eigenvalue λ_1 of the matrix $\mathbf{W}^{-1} \mathbf{B}$, which leads to what is called Roy's largest root test (remembering that eigenvalues are also called latent roots). The basis for using this statistic is the result that if the linear combination of the variables $X_1 - X_p$ that maximizes the ratio of the between-sample sum of squares to the within-sample sum of squares is found, then this maximum ratio equals λ_1. This, then, implies that this maximum eigenvalue should be a good statistic for testing whether the between-sample variation is significantly large and that there is, therefore, evidence that the samples being considered do not come from populations with the same mean vector. This approach is related to discriminant function analysis, which is the subject of Chapter 8. It may be important to know that what some computer programs call Roy's largest root statistic is $\lambda_1/(1-\lambda_1)$ rather than λ_1 itself. If in doubt, check the program documentation.

To assess whether λ_1 is significantly large, the exact probability of a value as large as the observed one can be calculated numerically, or an F distribution can be used to find a lower bound to the significance level, that is, the F-value is calculated and the true significance level is greater than the probability of obtaining a value this large or more. Users of computer packages should be aware of which of these alternatives is used if a significant result is obtained. This is because if the F distribution is used, then the value of λ_1 may not actually be significantly large at the chosen significance level. The F-value used is described in Table 4.4.

The third statistic often used to test whether the samples come from populations with the same mean vectors is Pillai's trace statistic. This can be written in terms of the eigenvalues $\lambda_1 - \lambda_p$ as

$$V = \sum_{i=1}^{p} \lambda_i/(1+\lambda_i) \tag{4.16}$$

Again, large values of this statistic provide evidence that the samples being considered come from populations with different mean vectors. An approximation for the significance level (the probability of obtaining a value as large as or larger than V if the samples are from populations with the same mean vector) is again provided in Table 4.4.

Table 4.4 Test statistics used to compare sample mean vectors with approximate F tests for evidence that the population values are not constant

Test	Statistic	F	df_1	df_2	Comment
Wilks' lambda	Λ	$\{(1 - \Lambda^{1/t})/\Lambda^{1/t}\}$ (df_2/df_1)	$p(m-1)$	$wt - df_1/2 + 1$	$w = n - 1 - (p+m)/2$ $t = [(df_1^2 - 4)/\{p^2 + (m-1)^2 - 5\}]^{1/2}$ If $df_1 = 2$, set $t = 1$
Roy's largest root	λ_1	$(df_2/df_1)\lambda_1$	D	$n - m - d - 1$	The significance level obtained is a lower bound $d = \max(p, m-1)$
Pillai's trace	$V = \sum_{i=1}^{p} \lambda_i/(1+\lambda_i)$	$(n - m - p + s)V/$ $\{d(s - V)\}$	sd	$s(n - m - p + s)$	$s = \min(p, m-1) =$ number of positive eigenvalues $d = \max(p, m-1)$
Lawes–Hotelling trace	$U = \sum_{i=1}^{p} \lambda_i$	$df_2 \, U/(s \, df_1)$	$s(2A + s + 1)$	$2(sB + 1)$	s is as for Pillai's trace $A = (|m - p - 1| - 1)/2$ $B = (n - m - p - 1)/2$

Note: It is assumed that there are p variables in m samples, with the jth the of size n_j, and a total sample size of $n = \Sigma n_j$. These are approximations for general p and m. Exact or better approximations are available for some special cases, and other approximations are also available. In all cases, the test statistic is transformed to the stated F value, and this is tested to see whether it is significantly large in comparison with the F distribution with df_1 and df_2 degrees of freedom. Chi-squared distribution approximations are also in common use, and tables of critical values are available (Kres, 1983).

Finally, the fourth statistic often used to test the null hypothesis of equal population mean vectors is the Lawes–Hotelling trace

$$U = \sum_{i=1}^{p} \lambda_i \qquad (4.17)$$

which is just the sum of the eigenvalues of the matrix $\mathbf{W}^{-1} \mathbf{B}$. Yet again, large values provide evidence against the null hypothesis, with an approximate F-test provided in Table 4.4.

Generally, the four tests just described can be expected to give similar significance levels, so that there is no real need to choose between them. They all involve the assumption that the distribution of the p variables is multivariate normal, with the same within-sample covariance matrix for all the m populations from which the samples are drawn. They are all also considered to be fairly robust if the sample sizes are equal or nearly so for the m samples. If there are questions about either the multivariate normality or the equality of covariance matrices, then simulation studies suggest that Pillai's trace statistic may be more robust than the other three statistics (Seber, 2004, p. 442).

4.8   Comparison of variation for several samples

Box's M-test is the best known for comparing the variation in several samples. This test has already been mentioned for the two-sample situations with several variables to be compared, and it can be used with one or several variables, with two or more samples.

For m samples, the M statistic is given by the equation

$$M = \left\{ \prod_{i=1}^{m} |C_i|^{(n_i-1)/2} \right\} \Big/ |C|^{(n-m)/2} \qquad (4.18)$$

where:

n_i	is the size of the ith sample
C_i	is the sample covariance for the ith sample as defined in Section 2.7
C	is the pooled covariance matrix

$$C = \sum_{i=1}^{m} (n_i - 1) \; C_i / (n - m)$$

$n = \sum n_i$ is the total number of observations

Large values of M provide evidence that the samples are not from populations with the same covariance matrix. An approximate F-test for whether an observed M-value is significantly large is provided by calculating

$$F = -2 \cdot b \log_e (M) \tag{4.19}$$

and finding the probability of a value this large or larger for an F distribution with v_1 and v_2 df, where

$$v_1 = p(p+1)(m-1)/2$$

$$v_2 = (v_1+2)/(c_2-c_1^2)$$

and

$$b = (1-c_1-v_1/v_2)/v_1$$

where

$$c_1 = (2p^2+3p-1)\left\{\sum_{i=1}^{m} 1/(n_i-1) - 1/(n-m)\right\}/\{6(p+1)(m-1)\}$$

and

$$c_2 = (p-1)(p+2)\left\{\sum_{i=1}^{m} 1/(n_i-1)^2 - 1(n-m)^2\right\}/\{6(m-1)\}$$

The F approximation of Equation 4.19 is only valid for $c_2 > c_1^2$. If $c_2 < c_1^2$, then an alternative approximation is used. In this alternative case, the F-value is calculated to be

$$F = -\{2\,b_1 \cdot v_2 \cdot \log_e (M)\}/\{v_1 + 2 \cdot b_1 \cdot \log_e (M)\} \tag{4.20}$$

where

$$b_1 = (1-c_1-2/v_2)/v_2$$

This is tested against the F distribution with v_1 and v_2 df to see whether it is significantly large.

Box's test is known to be sensitive to deviations from normality in the distribution of the variables being considered. For this reason, robust alternatives to Box's test are recommended here, these being generalizations of what was suggested for the two-sample situation. Thus, absolute deviations from sample medians can be calculated for the data in m samples. For a single variable, these can be treated as the observations for a one-factor analysis of variance. A significant F-ratio is then evidence that the samples come from populations with different mean deviations, that is, populations with different covariance matrices. With more than one variable, any of the four tests described in the last section can be applied to the transformed data, and a significant result indicates that the covariance matrix is not constant for the m populations sampled.

Alternatively, the variables can be standardized to have unit variances for all the data lumped together and d-values calculated using Equation 4.11. These d-values can then be analyzed by a one-factor analysis of variance. This generalizes Van Valen's test, which was suggested for comparing the variation in two multivariate samples. A significant F-ratio from the analysis of variance indicates that some of the m populations sampled are more variable than others. As in the two-sample situation, this test is only really appropriate when some samples may be more variable than others for all the measurements being considered.

Example 4.3: Comparison of samples of Egyptian skulls

As an example of the test for comparing several samples, consider the data shown in Table 1.2 for four measurements on male Egyptian skulls for five samples from various past ages.

A one-factor analysis of variance on the first variable, maximum breadth, provides $F = 5.95$, with 4 and 145 df (Table 4.3). This is significantly large at the 0.1% level, and hence there is clear evidence that the population mean changed with time. For the other three variables, analysis of variance provides the following results: basibregmatic height, $F = 2.45$ (significant at the 5% level); basialveolar length, $F = 8.31$ (significant at the 0.1% level); and nasal height, $F = 1.51$ (not significant). Hence, there is evidence that the population mean changed with time for the first three variables.

Next, consider the four variables together. If the five samples are combined, then the matrix of sums of squares and products for the 150 observations, calculated using Equation 4.13, is

$$\mathbf{T} = \begin{bmatrix} 3563.89 & -222.81 & -615.16 & 291.30 \\ -222.81 & 3635.17 & 1046.28 & 346.47 \\ -615.16 & 1046.28 & 4309.27 & -16.40 \\ 426.73 & 346.47 & -16.40 & 1533.33 \end{bmatrix}$$

for which the determinant is $\mathbf{ITI} = 7.306 \times 10^{13}$. Also, the within-sample matrix of sums of squares and cross products is found from Equation 4.14 to be

$$\mathbf{W} = \begin{bmatrix} 3061.07 & 5.33 & 11.47 & 291.30 \\ 5.33 & 3405.27 & 754.00 & 412.53 \\ 11.47 & 754.00 & 3505.97 & 164.33 \\ 291.30 & 412.53 & 164.33 & 1472.13 \end{bmatrix}$$

for which the determinant is $|\mathbf{W}| = 4.848 \times 10^{13}$. Wilks' lambda statistic is therefore

$$\Lambda = |\mathbf{W}| / |\mathbf{T}| = 0.6636$$

The details of an approximate F-test to assess whether this value is significantly small are provided in Table 4.4. With $p = 4$ variables, $m = 5$ samples, and $n = 150$ observations in total, it is found using the notation in Table 4.4 that

$$df_1 = p(m-1) = 16$$
$$w = n - 1 - (p+m)/2 = 150 - 1 - (4+5)/2 = 144.5$$
$$t = \left[\left(df_1^2 - 4 \right) / \left\{ p^2 + (m-1)^2 - 5 \right\} \right]^{\frac{1}{2}} = \left[\left(16^2 - 4 \right) / \left\{ 4^2 + (5-1)^2 - 5 \right\} \right]^{\frac{1}{2}}$$
$$= 3.055$$

and

$$df_2 = wt - df_1/2 + 1 = 144.5 \times 3.055 - 16/2 + 1 = 434.5$$

The F-statistic is then

$$F = \left\{ \left(1 - \Lambda^{1/t} \right) / \Lambda^{1/t} \right\} (df_2 / df_1) = \left\{ \left(1 - 0.6636^{1/3.055} \right) / 0.6636^{1/3.055} \right\}$$
$$(434.5 / 16) = 3.90$$

with 16 and 434.5 df. This is significantly large at the 0.1% level ($p < 0.001$). There is, therefore, clear evidence that the vector of mean values of the four variables changed with time.

The maximum root of the matrix $\mathbf{W}^{-1} \mathbf{B}$ is $\lambda_1 = 0.4251$ for Roy's maximum root test. The corresponding approximate F-statistic from Table 4.4 is

$$F = (df_2/df_1)\lambda_1 = (140/4)0.4251 = 14.88$$

with 4 and 140 df, using the equations given in the table for the df. This, again, is very significantly large (p < 0.001).

Pillai's trace statistic is V = 0.3533. The approximate F-statistic in this case is

$$F = (n - m - p + s)V/\{d(s - V)\} = 3.51$$

with s d = 16 and s (n − m − p + s) = 580 df, using the equations given in Table 4.4. This is another very significant result (p < 0.001).

Finally, for the tests on the mean vectors, the Lawley–Hotelling trace statistic has the value U = 0.4818. It is found using the equations in Table 4.4 that the intermediate quantities that are needed are s = 4, A = −0.5, and B = 70, so that the df for the F-statistic are $df_1 = s(2A + s + 1) = 16$ and $df_2 = 2(sB + 1) = 562$. The F-statistic is then

$$F = df_2 \cdot U/(s \cdot df_1) = (562 \times 0.4818)/(4 \times 16) = 4.23$$

Yet again, this is a very significant result (p < 0.001).

To compare the variation in the five samples, first consider Box's test. Equation 4.18 gives $M = 2.869 \times 10^{-11}$. The equations in the last section then give b = 0.0235,

$$F = -2 \cdot b \cdot \log_e(M) = 1.14,$$

with $v_1 = 40$ and $v_2 = 46{,}379$ df. This is not at all significantly large (p = 0.250), so this test gives no evidence that the covariance matrix changed with time.

Box's test is reasonable with this set of data, because body measurements tend to have distributions that are close to normal. However, robust tests can also be carried out. It is a straightforward matter to transform the data into absolute deviations from sample medians for Levene-type tests. Analysis of variance then shows no significant difference between the sample means of the transformed data for any of the four variables considered individually. Also, none of the multivariate tests summarized in Table 4.4 give a result that is anything like significant at the 5% level for all the variables taken together.

It appears, therefore, that although there is very strong evidence that mean values changed with time for the four variables being considered, there is no evidence at all that the variation changed.

4.9   Computer programs

The tests for multivariate normal data that are discussed in this chapter are readily available in standard statistical computer packages, although

Table 4.5 Values for nine mandible measurements for samples of five canine groups

	X_1	X_2	X_3	X_4	X_5	X_6	X_7	X_8	X_9	Sex
				Modern dogs from Thailand						
1	123	10.1	23	23	19	7.8	32	33	5.6	1
2	137	9.6	19	22	19	7.8	32	40	5.8	1
3	121	10.2	18	21	21	7.9	35	38	6.2	1
4	130	10.7	24	22	20	7.9	32	37	5.9	1
5	149	12.0	25	25	21	8.4	35	43	6.6	1
6	125	9.5	23	20	20	7.8	33	37	6.3	1
7	126	9.1	20	22	19	7.5	32	35	5.5	1
8	125	9.7	19	19	19	7.5	32	37	6.2	1
9	121	9.6	22	20	18	7.6	31	35	5.3	2
10	122	8.9	20	20	19	7.6	31	35	5.7	2
11	115	9.3	19	19	20	7.8	33	34	6.5	2
12	112	9.1	19	20	19	6.6	30	33	5.1	2
13	124	9.3	21	21	18	7.1	30	36	5.5	2
14	128	9.6	22	21	19	7.5	32	38	5.8	2
15	130	8.4	23	20	19	7.3	31	40	5.8	2
16	127	10.5	25	23	20	8.7	32	35	6.1	2
				Golden jackals						
1	120	8.2	18	17	18	7.0	32	35	5.2	1
2	107	7.9	17	17	20	7.0	32	34	5.3	1
3	110	8.1	18	16	19	7.1	31	32	4.7	1
4	116	8.5	20	18	18	7.1	32	33	4.7	1
5	114	8.2	19	18	19	7.9	32	33	5.1	1
6	111	8.5	19	16	18	7.1	30	33	5.0	1
7	113	8.5	17	18	19	7.1	30	34	4.6	1
8	117	8.7	20	17	18	7.0	30	34	5.2	1
9	114	9.4	21	19	19	7.5	31	35	5.3	1
10	112	8.2	19	17	19	6.8	30	34	5.1	1
11	110	8.5	18	17	19	7.0	31	33	4.9	2
12	111	7.7	20	18	18	6.7	30	32	4.5	2
13	107	7.2	17	16	17	6.0	28	35	4.7	2
14	108	8.2	18	16	17	6.5	29	33	4.8	2
15	110	7.3	19	15	17	6.1	30	33	4.5	2
16	105	8.3	19	17	17	6.5	29	32	4.5	2
17	107	8.4	18	17	18	6.2	29	31	4.3	2
18	106	7.8	19	18	18	6.2	31	32	4.4	2
19	111	8.4	17	16	18	7.0	30	34	4.7	2
20	111	7.6	19	17	18	6.5	30	35	4.6	2

(*Continued*)

Multivariate Statistical Methods: A Primer

Table 4.5 (Continued) Values for nine mandible measurements for samples of five canine groups

	X_1	X_2	X_3	X_4	X_5	X_6	X_7	X_8	X_9	Sex
					Cuons					
1	123	9.7	22	21	20	7.8	27	36	6.1	1
2	135	11.8	25	21	23	8.9	31	38	7.1	1
3	138	11.4	25	25	22	9.0	30	38	7.3	1
4	141	10.8	26	25	21	8.1	29	39	6.6	1
5	135	11.2	25	25	21	8.5	29	39	6.7	1
6	136	11.0	22	24	22	8.1	31	39	6.8	1
7	131	10.4	23	23	23	8.7	30	36	6.8	1
8	137	10.6	25	24	21	8.3	28	38	6.5	1
9	135	10.5	25	25	21	8.4	29	39	6.9	1
10	131	10.9	25	24	21	8.5	29	35	6.2	2
11	130	11.3	22	23	21	8.7	29	37	7.0	2
12	144	10.8	24	26	22	8.9	30	42	7.1	2
13	139	10.9	26	23	22	8.7	30	39	6.9	2
14	123	9.8	23	22	20	8.1	26	34	5.6	2
15	137	11.3	27	26	23	8.7	30	39	6.5	2
16	128	10.0	22	23	22	8.7	29	37	6.6	2
17	122	9.9	22	22	20	8.2	26	36	5.7	2
					Indian wolves					
1	167	11.5	29	28	25	9.5	41	45	7.2	1
2	164	12.3	27	26	25	10.0	42	47	7.9	1
3	150	11.5	21	24	25	9.3	41	46	8.5	1
4	145	11.3	28	24	24	9.2	36	41	7.2	1
5	177	12.4	31	27	27	10.5	43	50	7.9	1
6	166	13.4	32	27	26	9.5	40	47	7.3	1
7	164	12.1	27	24	25	9.9	42	45	8.3	1
8	165	12.6	30	26	25	7.7	40	43	7.9	1
9	131	11.8	20	24	23	8.8	38	40	6.5	2
10	163	10.8	27	24	24	9.2	39	48	7.0	2
11	164	10.7	24	23	26	9.5	43	47	7.6	2
12	141	10.4	20	23	23	8.9	38	43	6.0	2
13	148	10.6	26	21	24	8.9	39	40	7.0	2
14	158	10.7	25	25	24	9.8	41	45	7.4	2
					Prehistoric Thai dogs					
1	112	10.1	17	18	19	7.7	31	33	5.8	0
2	115	10.0	18	23	20	7.8	33	36	6.0	0

(Continued)

Table 4.5 (Continued) Values for nine mandible measurements for
samples of five canine groups

	X_1	X_2	X_3	X_4	X_5	X_6	X_7	X_8	X_9	Sex
3	136	11.9	22	25	21	8.5	36	39	7.0	0
4	111	9.9	19	20	18	7.3	29	34	5.3	0
5	130	11.2	23	27	20	9.1	35	35	6.6	0
6	125	10.7	19	26	20	8.4	33	37	6.3	0
7	132	9.6	19	20	19	9.7	35	38	6.6	0
8	121	10.7	21	23	19	7.9	32	35	6.0	0
9	122	9.8	22	23	18	7.9	32	35	6.1	0
10	124	9.5	20	24	19	7.6	32	37	6.0	0

Note: The variables are X_1 = length of mandible, X_2 = breadth of mandible below 1st molar, X_3 = breadth of articular condyle, X_4 = height of mandible below first molar, X_5 length of first molar, X_6 = breadth of first molar, X_7 = length of first to third molar inclusive (first to second for cuon), X_8 = length from first to fourth premolar inclusive, and X_9 = breadth of lower canine, all measured in millimeters. The Sex code is 1 for male, 2 for female, and 0 for unknown.

many packages will be missing one or two of them. The tests can also be carried out using the R code described in the Appendix to this chapter. The results of tests based on F distribution approximations may vary to some extent from one program to the next because of the use of different approximations. On the other hand, for the robust tests on variances, some or all of the calculations may need to be done in a spreadsheet.

This chapter has been restricted to situations in which there are two or more multivariate samples being compared to see whether they seem to come from populations with different mean vectors or from populations with different covariance matrices. In terms of mean vectors, this is the simplest case of what is sometimes called *multivariate analysis of variance* (MANOVA). More complicated examples involve samples being classified on the basis of several factors, giving a generalization of ordinary analysis of variance (ANOVA). Most statistical packages allow the general MANOVA calculations to be performed.

Exercise

Example 1.4 concerned the comparison between prehistoric dogs from Thailand and six other related animal groups in terms of mean mandible measurements. Table 4.5 shows some further data for the comparison of these groups, which are part of the more extensive data discussed in the paper by Higham et al. (1980).

1. Test for significant differences between the five species in terms of the mean values and the variation in the nine variables. Test both for overall differences and for differences between the

prehistoric Thai dogs and each of the other groups singly. What conclusion do you draw with regard to the similarity between prehistoric Thai dogs and the other groups?

2. Is there evidence of differences between the size of males and females of the same species for the first four groups?

3. Using a suitable graphical method, compare the distribution of the nine variables for the prehistoric and modern Thai dogs.

References

Higham, C.F.W., Kijngam, A. and Manly, B.F.J. (1980). An analysis of prehistoric canid remains from Thailand. *Journal of Archaeological Science* 7: 149–65.

Kres, H. (1983). *Statistical Tables for Multivariate Analysis*, New York: Springer.

Levene, H. (1960). Robust tests for equality of variance. In *Contributions to Probability and Statistics* (eds I. Olkin, S.G. Ghurye, W. Hoeffding, W.G. Madow and H.B. Mann), pp. 278–92. Pala Alto, CA: Stanford University Press.

Manly, B.F.J. (2009). *Statistics for Environmental Science and Management*. 2nd Edn. Boca Raton, FL: Chapman and Hall/CRC.

Manly, B.F.J. and Francis, R.I.C.C. (2002). Testing for mean and variance differences with samples from distributions that may be non-normal with unequal variances. *Journal of Statistical Computation and Simulation* 72: 633–46.

Schultz, B. (1983). On Levene's test and other statistics of variation. *Evolutionary Theory* 6: 197–203.

Seber, G.A.F. (2004). *Multivariate Observations*. New York: Wiley.

Van Valen, L. (1978). The statistics of variation. *Evolutionary Theory* 4: 33–43. (Erratum *Evolutionary Theory* 4: 202.)

Welch, B.L. (1951). On the comparison of several mean values: An alternative approach. *Biometrika* 38: 330–6.

Yao, Y. (1965). An approximate degrees of freedom solution to the multivariate Behrens-Fisher problem. *Biometrika* 52: 139–47.

Appendix: Tests of Significance in R

A.1   Univariate two-sample t-test in R

The calculations for the two-sample t-test are achieved with the `t.test()` function. By default, R assumes a two-sided test and that the within-sample variances are not equal. Therefore, under the assumption of common variances, it is necessary to write

```
t.test(vector1, vector2, var.equal = TRUE)
```

where `vector1` and `vector2` are the numeric vectors for Samples 1 and 2 of a single response variable, respectively. The supplementary material provided at the book's website shows the comparison of the mean total length between survivor and nonsurvivor Bumpus' sparrows.

A.2   Multivariate two-sample tests with Hotelling's T^2

The command with the name `hotelling.test()` can be executed after loading the package `Hotelling` (Curran et al. 2013), using

```
library(Hotelling)
hotelling.test(formula)
```

Here, `formula` is an expression in which the response variables, separated by sum (+) signs, are followed by a tilde (~) and a factor vector with two levels. As an example, if `Y1`, `Y2`, and `Y3` are three response variables in a data frame, and `X` is a two-level grouping variable, then `hotelling.test(Y1, Y2, Y3 ~ X)` invokes Hotelling's test for the comparison of the two multivariate means. A script with this function to get the results of Example 4.1 can be found at the book's website.

A.3   Comparison of variation of two samples
     for the univariate case

R offers the two-sample variance tests described in Section 4.5 using the F-ratio test and Levene's test, where the latter requires the package `car` to be loaded. The F-test is invoked via the `var.text` function, its syntax being similar to the `t.test()` function:

```
var.test(vector1, vector2)
```

By default, the alternative hypothesis for this test is two-sided.

Levene's test, as implemented in package `car` (the `leveneTest()` function), does not restrict the grouping variable to two levels only. The code for this function is

```
library(car)
leveneTest(response ~ x)
```

The first argument in the `leveneTest()` function is a vector containing the response variable, and x is the factor defining groups (with at least two levels).

A.4   Comparison of variation of two multivariate samples

The R commands needed for testing the multivariate variation of two groups as outlined in Section 4.6 make use of vector functions as well as a function for data standardization, either `scale()` or `decostand()`. While `scale()` has a simpler syntax, `decostand()` (a function from the package vegan (Oksanen et al. 2016)) is more versatile, but it requires the numeric (column) data vectors to be put into matrix form. This book's website provides the scripts needed for running the two procedures applied to the Bumpus' sparrows data for Example 4.2. These are the Levene-type test, based on the T^2-statistic, and Van Valen's test, based on the univariate two-sample t-test.

A.5   Comparison of several multivariate means

R offers the function `manova()` for MANOVA, which is an extension of the corresponding univariate R-function aov. In fact, `manova()` calls `aov()` for computations, but supplementary information is also computed and made available as output, including sums of squares and cross products matrices, eigenvalues, and statistics such as Wilks' lambda and Pillai's test. This is described by saying that `manova()` produces a different object, one of class manova (see more details in the R documentation). The simplest call to a one-way MANOVA in R is

```
manova.obj <- manova(mat ~ x, data = df)
```

Here, the analyzed response variables are contained in matrix mat, taken from the data frame df, and the samples to be compared are levels of the factor x. Once the output of the `manova()` function is assigned to the object manova.obj, any of the test statistics available as summaries of the MANOVA can be invoked with `summary.manova()`. As an example, the results of a Wilks' test can be obtained with the instruction

```
summary.manova(manova.obj, test = "Wilks")
```

Options for the `test=` argument are "Wilks", "Pillai" (the default), "Hotelling-Lawley", and "Roy". The `manova` and `summary.manova` functions are applied to the comparison of samples of Egyptian skulls in the script that is available at this book's website.

A.6   Testing the equality of several covariance matrices

Box's M-test is carried out using the function `BoxM()` found in a package called `biotools` (da Silva, 2015). This function uses a chi-squared approximation instead of the procedure described in Section 4.8, which is based on the F distribution. Morrison (2004) gives the computational details for this chi-squared approximation, which is applicable whenever the number of groups and the number of variables do not exceed four or five, and the sample sizes per group are equal to or greater than 20. The F approximation is more suitable for greater numbers of groups and variables and smaller sample sizes per group. The formulae for both approximations are linked, so that it is not difficult to obtain the approximated F-statistic from the output generated by the function `BoxM()`. An example of the calculations of these two tests, used in Example 4.3, is available in this book's website.

References

Curran, J.M. (2013). Hotelling: Hotelling's T-squared test and variants. R package version 1.0-2. http://CRAN.R-project.org/package=Hotelling

da Silva, A.R. (2015). biotools: Tools for Biometry and Applied Statistics in Agricultural Science. R package version 2.1. http://CRAN.R-project.org/package=biotools

Morrison, D. (2004) *Multivariate Statistical Methods.* 4th Edn. Pacific Grove, CA: Duxbury.

Oksanen, J., Blanchet, F.G., Friendly, M., Kindt, R., Legendre, P., McGlinn, D., Minchin, P.R., et al. (2016). vegan: Community ecology package. R package version 2.4-0. http://CRAN.R-project.org/package=vegan

chapter five

Measuring and testing multivariate distances

5.1 Multivariate distances

A large number of multivariate problems can be viewed in terms of distances between single observations, between samples of observations, or between populations of observations. For example, considering the data in Table 1.4 on mandible measurements of dogs, wolves, jackals, cuons, and dingos, it is reasonable to ask how far one of these groups is from the other six groups. The idea, then, is that if two animals have similar mean mandible measurements, then they are close, whereas if they have rather different mean measurements, then they are distant from each other. Throughout this chapter, it is this concept of distance that is used.

A large number of distance measures have been proposed and used in multivariate analyses. Here, only some of the most common ones will be mentioned. It is fair to say that measuring distances is a topic in which a certain amount of arbitrariness seems unavoidable.

One situation is that there are n objects being considered, with a number of measurements being taken on each of these, and the measurements are of two types. For example, in Table 1.3, results are given for four environmental variables and six gene frequencies for 16 colonies of a butterfly. Two sets of distances, environmental and genetic, can therefore be calculated between the colonies. An interesting question then is whether there is a significant relationship between these two sets of distances. Mantel's test (Section 5.6) is useful in this context.

5.2 Distances between individual observations

To begin, consider the simplest case where there are n objects, each of which has values for p variables, $X_1, X_2, \dots, X_p$. The values for object i can then be denoted by $x_{i1}, x_{i2}, \dots, x_{ip}$, and those for object j by $x_{j1}, x_{j2}, \dots, x_{jp}$. The problem is to measure the distance between these two objects. If there are only $p = 2$ variables, then the values can be plotted as shown in Figure 5.1. Pythagoras' theorem then says that the length d_{ij} of the line joining the point for object i to the point for object j (the Euclidean distance) is

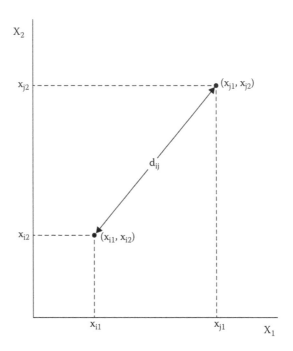

Figure 5.1 The Euclidean distance between objects i and j with p = 2 variables.

$$d_{ij} = \left\{ \left(x_{i1} - x_{j1} \right)^2 + \left(x_{i2} - x_{j2} \right)^2 \right\}^{\frac{1}{2}}$$

With $p = 3$ variables, the values can be taken as the coordinates in space for plotting the positions of individuals i and j (Figure 5.2). Pythagoras' theorem then gives the distance between the two points to be

$$d_{ij} = \left\{ \left(x_{i1} - x_{j1} \right)^2 + \left(x_{i2} - x_{j2} \right)^2 + \left(x_{i3} - x_{j3} \right)^2 \right\}^{\frac{1}{2}}$$

With more than three variables, it is not possible to use variable values as the coordinates for physically plotting points. However, the two- and three-variable cases suggest that the generalized Euclidean distance

$$d_{ij} = \left\{ \sum_{k=1}^{p} \left(x_{ik} - x_{jk} \right)^2 \right\} \tag{5.1}$$

may serve as a satisfactory measure for many purposes with p variables.

From the form of Equation 5.1, it is clear that if one of the variables measured is much more variable than the others, then this will dominate

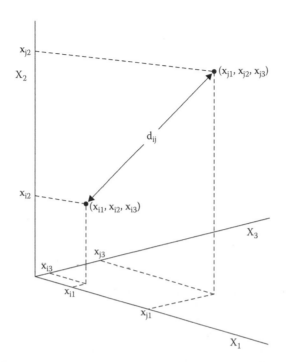

Figure 5.2 The Euclidean distance between objects i and j with $p = 3$ variables.

the calculation of distances. For example, to take an extreme case, suppose that n men are being compared and that X_1 is their stature and the other variables are tooth dimensions, with all the measurements being in millimeters. Stature differences will then be in the order of perhaps 20 or 30 mm, while tooth dimension differences will be in the order of 1 or 2 mm. The simple calculations of d_{ij} will then provide distances between individuals that are essentially stature differences only, with tooth differences having negligible effects.

In practice, it is usually desirable for all variables to have about the same influence on the distance calculation. This can be achieved by a preliminary scaling that involves dividing each variable by its standard deviation for the n individuals being compared.

Example 5.1: Distances between dogs and related species

Consider again the data in Table 1.4 for mean mandible measurements of seven groups of Thai dogs and related species. It may be recalled from Chapter 1 that the main question with these data is how the prehistoric Thai dogs relate to the other groups.

The first step in calculating distances is to standardize the measurements. Here, this will be done by expressing them as

Table 5.1 Standardized variable values calculated from the
original data in Table 1.4

Group	X_1	X_2	X_3	X_4	X_5	X_6
Modern dog	−0.46	−0.46	−0.68	−0.69	−0.45	−0.57
Golden jackal	−1.41	−1.79	−1.04	−1.29	−0.80	−1.21
Chinese wolf	1.78	1.48	1.70	1.80	1.55	1.50
Indian wolf	0.60	0.55	0.96	0.69	1.17	0.88
Cuon	0.13	0.31	−0.04	0.00	−1.10	−0.37
Dingo	−0.52	0.03	−0.13	−0.17	0.03	0.61
Prehistoric dog	−0.11	−0.12	−0.78	−0.34	−0.41	−0.83

deviations from means in units of standard deviations. For example,
the first measurement X_1 (breadth) has a mean of 10.486 mm and a
sample standard deviation of 1.697 mm for the seven groups. The
standardized variable values are then calculated as: modern dog,
$(9.7 - 10.486)/1.697 = -0.46$; golden jackal, $(8.1 - 10.486)/1.697 = -1.41$,
... prehistoric dog, $(10.3 - 10.486)/1.697 = -0.11$. Standardized values
for all the variables are shown in Table 5.1.

Using Equation 5.1, the distances shown in Table 5.2 have been calculated from the standardized variables. It is clear that the prehistoric
dogs are rather similar to modern dogs in Thailand, because the distance of 0.66 between these two groups is by far the smallest distance in
the whole table. Higham et al. (1980) concluded from a more complicated
analysis that the modern and prehistoric dogs are indistinguishable.

5.3   Distances between populations and samples

A number of measures have been proposed for the distance between multivariate populations when information is available on the means, variances, and covariances of the populations. Here, two measures will be
considered.

Suppose that two or more populations are available, and the
multivariate distributions in these populations are known for p variables

Table 5.2 Euclidean distances between seven canine groups

	Modern dog	Golden jackal	Chinese wolf	Indian wolf	Cuon	Dingo	Prehistoric dog
Modern dog	—						
Golden jackal	1.91	—					
Chinese wolf	5.38	7.12	—				
Indian wolf	3.38	5.06	2.14	—			
Cuon	1.51	3.19	4.57	2.91	—		
Dingo	1.56	3.18	4.21	2.20	1.67	—	
Prehistoric dog	0.66	2.39	5.12	3.24	1.26	1.71	—

$X_1, X_2, \ldots, X_p$. Let the mean of variable X_k in the ith population be μ_{ki}, and assume that the variance of X_k is V_k in all the populations. Then, Penrose (1953) proposed the relatively simple measure

$$P_{ij} = \left\{ \sum_{k=1}^{p} (\mu_{ki} - \mu_{kj})^2 / (p \cdot V_k) \right\}$$

(5.2)

for the distance between population i and population j.

A disadvantage of Penrose's measure is that it does not take into account the correlations between the p variables. This means that when two variables are measuring essentially the same thing, and hence are highly correlated, they still individually both contribute about the same amount to population distances as a third variable that is uncorrelated with all other variables.

A measure that does take into account the correlations between variables is the Mahalanobis (1948) distance

$$D_{ij}^2 = \left\{ \sum_{r=1}^{p} \sum_{s=1}^{p} (\mu_{ri} - \mu_{rj}) \cdot v^{rs} \cdot (\mu_{si} - \mu_{sj}) \right\},$$

(5.3)

where v^{rs} is the element in the rth row and sth column of the inverse of the population covariance matrix for the p variables. This is a quadratic form that can also be written as

$$D_{ij}^2 = (\mu_i - \mu_j)' V^{-1} (\mu_i - \mu_j),$$

where:

μ_i is the population mean vector for the ith population
V is the population covariance matrix

This measure does require the assumption that V is the same for all populations.

A Mahalanobis distance is also often used to measure the distance of a single multivariate observation from the center of the population that the observation comes from. If $x_1, x_2, \ldots, x_p$ are the values of $X_1, X_2, \ldots, X_p$ for the individual, with corresponding population mean values of $\mu_1, \mu_2, \ldots, \mu_p$, then this distance is

$$D^2 = \left\{ \sum_{r=1}^{p} \sum_{s=1}^{p} (x_r - \mu_r) \cdot v^{rs} \cdot (x_s - \mu_s) \right\}$$

$$= (x - \mu)' V^{-1} (x - \mu)$$

(5.4)

where:

$\quad \mathbf{x} \quad = (x_1, x_2, \ldots x_p)$
$\quad \boldsymbol{\mu} \quad$ is the population mean vector
$\quad \mathbf{V} \quad$ is the population covariance matrix
$\quad v^{rs} \quad$ is the element in the rth row and sth column of the inverse of $\mathbf{V}$

The value of D^2 can be thought of as a multivariate residual for the observation $\mathbf{x}$, that is, a measure of how far the observation $\mathbf{x}$ is from the center of the distributions of all values, taking into account all the variables being considered and their covariances. An important result is that if the population being considered is multivariate normally distributed, then the values of D^2 will follow a chi-squared distribution with p degrees of freedom (df) if $\mathbf{x}$ comes from this distribution. A significantly large value of D^2 means that the corresponding observation is either (a) a genuine but unlikely record, (b) an observation from another distribution, or (c) a record containing some mistake. Observations with large Mahalanobis residuals should therefore be examined to see whether they have just been recorded wrongly.

Equations 5.2 through 5.4 can be used with sample data if estimates of population means, variances, and covariances are used in place of true values. In that case, the covariance matrix $\mathbf{V}$ involved in Equations 5.3 and 5.4 should be replaced with the pooled estimate from all the samples available, as defined in Section 4.8 for Box's M-test.

In principle the Mahalanobis distance is superior to the Penrose distance, because it uses information on covariances. However, this advantage is only present when the covariances are accurately known. When covariances can only be estimated rather poorly from small samples, it is probably best to use the simpler Penrose measure. It is difficult to say precisely what a small sample means in this context. Certainly, there should be no problem with using Mahalanobis distances based on a covariance matrix estimated with a total sample size of 100 or more.

Example 5.2: Distances between samples of Egyptian skulls

For the five samples of male Egyptian skulls shown in Table 1.2, the mean vectors and covariance matrices are shown in Table 5.3, as is the pooled covariance matrix. Although the five sample covariance matrices appear to differ somewhat, it has been shown in Example 4.3 that the differences are not significant.

Penrose's distance measures of Equation 5.2 can now be calculated between each pair of samples. There are p = 4 variables with variances that are estimated by $V_1 = 21.112$, $V_2 = 23.486$, $V_3 = 24.180$, and $V_4 = 10.154$, these being the diagonal terms in the

Table 5.3 The samples' mean vectors and covariance matrices and the pooled sample covariance matrix for the Egyptian skull data

Sample		Mean vector	Sample covariance matrices			
			X_1	X_2	X_3	X_4
1	X_1	131.37	26.31	4.15	0.45	7.25
	X_2	133.60	4.15	19.97	−0.79	0.39
	X_3	99.17	0.45	−0.79	34.63	−1.92
	X_4	50.53	7.25	0.39	−1.92	7.64
2	X_1	132.37	23.14	1.01	4.77	1.84
	X_2	132.70	1.01	21.60	3.37	5.62
	X_3	99.07	4.77	3.37	18.89	0.19
	X_4	50.23	1.84	5.62	0.19	8.74
3	X_1	134.47	12.12	0.79	−0.78	0.90
	X_2	133.80	0.79	24.79	3.59	−0.09
	X_3	96.03	−0.78	3.59	20.72	1.67
	X_4	50.57	0.90	−0.09	1.67	12.60
4	X_1	135.50	15.36	−5.53	−2.17	2.05
	X_2	132.30	−5.53	26.36	8.11	6.15
	X_3	94.53	−2.17	8.11	21.09	5.33
	X_4	51.97	2.05	6.15	5.33	7.96
5	X_1	136.17	28.63	−0.23	−1.88	−1.99
	X_2	130.33	−0.23	24.71	11.72	2.15
	X_3	93.50	−1.88	11.72	25.57	0.40
	X_4	51.37	−1.99	2.15	0.40	13.83
			Pooled covariance matrix			
			21.112	0.038	0.078	2.010
			0.038	23.486	5.200	2.844
			0.078	5.200	24.180	1.134
			2.010	2.844	1.134	10.154

pooled covariance matrix (Table 5.3). The sample mean values given in the vectors x_1 to x_5 are estimates of population means. For example, the distance between Sample 1 and Sample 2 is calculated as

$$P_{12} = (131.37 - 132.37)^2 / (4 \times 21.112) + (133.60 - 132.70)^2 / (4 \times 23.486)$$

$$+ (99.17 - 99.07)^2 / (4 \times 24.180) + (50.53 - 50.23)^2 / (4 \times 10.154)$$

$$= 0.023$$

This only has meaning in comparison with the distances between the other pairs of samples. Calculating these as well provides the distances shown in Table 5.4a.

It may be recalled from Example 4.3 that the mean values change significantly from sample to sample. The Penrose distances show that the changes are cumulative over time, with the samples that are closest in time being relatively similar, whereas the samples that are far apart in time are more different.

Turning next to Mahalanobis distances, these can be calculated from Equation 5.3, with the population covariance matrix **V** estimated by the pooled sample covariance matrix **C**. The matrix **C** is provided in Table 5.3, and the inverse is

$$\mathbf{C}^{-1} = \begin{bmatrix} 0.0483 & 0.0011 & 0.0001 & -0.0099 \\ 0.0011 & 0.0461 & -0.0094 & -0.0121 \\ 0.0001 & -0.0094 & 0.0435 & -0.0022 \\ -0.0099 & -0.0121 & -0.0022 & 0.1041 \end{bmatrix}$$

Using this inverse and the sample means gives the Mahalanobis distance from Sample 1 to Sample 2 to be

Table 5.4 Penrose and Mahalanobis distances between pairs of samples of Egyptian skulls

	Early predynastic	Late predynastic	12th–13th Dynasties	Ptolemaic	Roman
(a) Penrose distances					
Early predynastic	—				
Late predynastic	0.023	—			
12th–13th dynasties	0.216	0.163	—		
Ptolemaic	0.493	0.404	0.108	—	
Roman	0.736	0.583	0.244	0.066	—
(b) Mahalanobis distance					
Early predynastic	—				
Late predynastic	0.091	—			
12–13th dynasties	0.903	0.729	—		
Ptolemaic	1.881	1.594	0.443	—	
Roman	2.697	2.176	0.911	0.219	—

$$D_{12}^2 = (131.37 - 132.37)0.0483(131.37 - 132.37)$$
$$+ (131.37 - 132.37)0.0011(133.60 - 132.70)$$
$$+ ... - (50.53 - 50.23)0.0022(99.17 - 99.07)$$
$$+ (50.53 - 50.23)0.1041(50.53 - 50.23)$$
$$= 0.091$$

Calculating the other distances between samples in the same way provides the distance matrix shown in Table 5.4b.

A comparison between these distances and the Penrose distances shows a very good agreement. The Mahalanobis distances are three to four times as large as the Penrose distances. However, the relative distances between samples are almost the same for both measures. For example, the Penrose measure suggests that the distance from the early predynastic sample to the Roman sample is $0.736/0.023 = 32.0$ times as great as the distance from the early predynastic to the late predynastic sample. The corresponding ratio for the Mahalanobis measure is $2.697/0.091 = 29.6$.

5.4 Distances based on proportions

A particular situation that sometimes occurs is that the variables being used to measure the distance between populations or samples are proportions that add to one. For example, the animals of a certain species might be classified into K genetic classes. One colony might then have proportions p_1 of Class 1, p_2 of Class 2, up to p_K of Class K, while a second colony has proportions q_1 of Class 1, q_2 of Class 2, up to q_K of Class K. The question then arises of how different the extent of the genetic difference is between the two colonies.

Various indices of distance have been proposed with this type of proportion data. For example,

$$d_1 = \sum_{i=1}^{K} |p_i - q_i| / 2 \tag{5.5}$$

which is half of the sum of absolute proportion differences, is one possibility. This index takes the value of one when there is no overlap of classes and the value zero when $p_i = q_i$ for all i. Another possibility is

$$d_2 = 1 - \sum_{i=1}^{K} p_i\, q_i \left/ \left\{ \sum_{i=1}^{K} p_i^2 \sum q_i^2 \right\}^{1/2} \right. \qquad (5.6)$$

which again varies from one (no overlap) to zero (equal proportions).

Because d_1 and d_2 vary from zero to one, it follows that $1 - d_1$ and $1 - d_2$ are measures of the similarity between the cases being compared. In fact, it is in terms of similarities that the indices are often used. For example,

$$s_1 = 1 - d_1 = \sum |p_i - q_i| / 2$$

is often used as a measure of the niche overlap between two species, where p_i is the fraction of the resources used by Species 1 that are of Type i and q_i is the fraction of the resources used by Species 2 that are of Type i. Then, $s_1 = 0$ indicates that the two species use completely different resources, and $s_1 = 1$ indicates that the two species use exactly the same resources.

A similarity measure can also be constructed from any distance measure D that varies from zero to infinity. Taking $S = 1/D$ gives a similarity that ranges from infinity for two items that are no distance apart, to zero for two objects that are infinitely far apart. Alternatively, $1/(1 + D)$ ranges from 1 when $D = 0$, to 0 when D is infinite.

5.5 Presence-absence data

Another common situation is where the similarity or distance between two items must be based on a list of their presences and absences. For example, there might be interest in the similarity between two plant species in terms of their distributions at ten sites. The data might then take the form shown in Table 5.5. Such data are often summarized, as shown in Table 5.6, as counts of the number of times that both species are present (a), only one species is present (b and c), or both species are absent (d). Thus for the data in Table 5.5, $a = 3$, $b = 3$, $c = 3$, and $d = 1$.

Table 5.5 Presences and absences of two species at 10 sites (1 = presence, 0 = absence)

Site	1	2	3	4	5	6	7	8	9	10
Species 1	0	0	1	1	1	0	1	1	1	0
Species 2	1	1	1	1	0	0	0	0	1	1

Table 5.6 Presence and absence data obtained for
two species at n sites

Species 1	Species 2		
	Present	Absent	Total
Present	a	b	a+b
Absent	c	d	c+d
Total	a+c	b+d	N

In this situation, some of the commonly used similarity measures are

$$\text{the simple matching index} = (a+d)/n$$

$$\text{the Ochiai index} = a / \{(a+b)(a+c)\}^{\frac{1}{2}}$$

$$\text{the Dice–Sorensen index} = 2a / (2a+b+c)$$

and

$$\text{the Jaccard index} = a/(a+b+c)$$

These all vary from zero (no similarity) to one (complete similarity), so that complementary distance measures can be calculated by subtracting the similarity indices from one. These and other indices are reviewed by Gower and Legendre (1986), while Jackson et al. (1989) compare the results of using different indices with various multivariate analyses of the presences and absences of 25 fish species in 52 lakes.

There has been some debate about whether the number of joint absences (d) should be used in the calculation because of the danger of concluding that two species are similar simply because they are both absent from many sites. This is certainly a valid point in many situations, and suggests that the simple matching index should be used with caution.

5.6   The Mantel randomization test

A useful test for comparing two distance or similarity matrices was introduced by Mantel (1967) as a solution to the problem of detecting space and time clustering of diseases, that is, whether cases of a disease that occur close in space also tend to be close in time.

To understand the nature of the procedure, the following simple example should be helpful. Suppose that four objects are being studied, and that two sets of variables have been measured for each of these. The first set of variables can then be used to construct a 4×4 matrix, where the entry in the ith row and jth column is a measure of the distance between object i and object j based on these variables. The distance matrix might, for example, be

$$\mathbf{M} = \begin{bmatrix} m_{11} & m_{12} & m_{13} & m_{14} \\ m_{21} & m_{22} & m_{23} & m_{24} \\ m_{31} & m_{32} & m_{33} & m_{34} \\ m_{41} & m_{42} & m_{43} & m_{44} \end{bmatrix} = \begin{bmatrix} 0.0 & 1.0 & 1.4 & 0.9 \\ 1.0 & 0.0 & 1.1 & 1.6 \\ 1.4 & 1.1 & 0.0 & 0.7 \\ 0.9 & 1.6 & 0.7 & 0.0 \end{bmatrix}$$

This is a symmetric matrix, because, for example, the distance from Object 2 to Object 3 must be the same as the distance from Object 3 to Object 2 (1.1 units). Diagonal elements are zero, because these represent distances from objects to themselves.

The second set of variables can also be used to construct a matrix of distances between the objects. For the example, this will be taken as

$$\mathbf{E} = \begin{bmatrix} e_{11} & e_{12} & e_{13} & e_{14} \\ e_{21} & e_{22} & e_{23} & e_{24} \\ e_{31} & e_{32} & e_{33} & e_{34} \\ e_{41} & e_{42} & e_{43} & e_{44} \end{bmatrix} = \begin{bmatrix} 0.0 & 0.5 & 0.8 & 0.6 \\ 0.5 & 0.0 & 0.5 & 0.9 \\ 0.8 & 0.5 & 0.0 & 0.4 \\ 0.6 & 0.9 & 0.4 & 0.0 \end{bmatrix}$$

Like $\mathbf{M}$, this is symmetric with zeros down the diagonal.

Mantel's test is concerned with assessing whether the elements in $\mathbf{M}$ and $\mathbf{E}$ show some significant correlation. The test statistic that is used is sometimes the correlation between the corresponding elements of the two matrices (matching m_{11} with e_{11}, m_{12} with e_{12}, etc.), or the simpler sum of the products of these matched elements. For the general case of $n \times n$ matrices, the latter statistic is then

$$Z = \sum_{i=2}^{n} \left\{ \sum_{j=1}^{i-1} m_{ij} \cdot e_{ij} \right\} \tag{5.7}$$

This statistic is calculated and compared with the distribution of Z that is obtained by taking the objects in a random order for one of the matrices, which is why it is called a *randomization test*.

For the randomization test, the matrix $\mathbf{M}$ can be left as it is. A random order can then be chosen for the objects for matrix $\mathbf{E}$. For example, suppose that a random ordering of objects turns out to be 3,2,4,1. This then gives a randomized $\mathbf{E}$ matrix of

$$\mathbf{E}_R = \begin{bmatrix} 0.0 & 0.5 & 0.4 & 0.8 \\ 0.5 & 0.0 & 0.9 & 0.5 \\ 0.4 & 0.9 & 0.0 & 0.6 \\ 0.8 & 0.5 & 0.6 & 0.0 \end{bmatrix}$$

The entry in row 1 and column 2 is 0.5, the distance between objects 3 and 2, the entry in row 1 and column 3 is 0.4, the distance between objects 3 and 4, and so on. A Z value can be calculated using $\mathbf{M}$ and $\mathbf{E}_R$. Repeating this procedure using different random orders of the objects for $\mathbf{E}_R$ generates the randomized distribution of Z. A check can then be made to see whether the observed Z value is a typical value from this distribution.

The basic idea here is that if the two measures of distance are quite unrelated, then the matrix $\mathbf{E}$ will be just like one of the randomly ordered matrices $\mathbf{E}_R$. Hence, the observed Z will be a typical randomized Z value. On the other hand, if the two distance measures have a positive correlation, then the observed Z will tend to be larger than values given by randomization. A negative correlation between distances should not occur, but if it does, then the result will be that the observed Z value will tend to be low when compared with the randomized distribution.

With n objects, there are n! different possible orderings of the object numbers. There are therefore n! possible randomizations of the elements of $\mathbf{E}$, some of which might give the same Z values. Hence, in our example with four objects, the randomized Z-distribution has $4! = 24$ equally likely values. It is not too difficult to calculate all these. More realistic cases might involve, say, 15 objects, in which case the number of possible Z values is $15! \approx 1.3 \times 10^{12}$. Enumerating all of these then becomes impractical, and there are two possible approaches for carrying out the Mantel test. A large number of randomized $\mathbf{E}_R$ matrices can be generated on a computer and the resulting distribution of Z values used in place of the true randomized distribution. Alternatively, the mean, E(Z), and variance, Var(Z), of the randomized distribution of Z can be calculated, and

$$g = \{Z - E(Z)\} / \mathrm{Var}(Z)$$

can be treated as a standard normal variate. Mantel (1967) provided formulae for the mean and variance of Z in the null hypothesis case of no correlation between the distance measures. There is, however, some doubt about the validity of the normal approximation for the test statistic q (Mielke, 1978), so that it seems best to perform randomizations rather than to rely on this approximation.

The test statistic Z of Equation 5.7 is the sum of the products of the elements in the lower diagonal parts of the matrices **M** and **E**. The only reason for using this particular statistic is that Mantel's equations for the mean and variance are available. However, if it is decided to determine significance by randomizations, then there is no reason why the test statistic should not be changed. Indeed, values of Z are not particularly informative except in comparison with the randomization mean and variance. It may, therefore, be more useful to take the correlation r_{ME} between the lower diagonal elements of **M** and **E** as the test statistic instead of Z. With $n \times n$ matrices, there are $n(n-1)/2$ lower diagonal terms, which pair up as (m_{21}, e_{21}), (m_{31}, e_{31}), (m_{32}, e_{32}), and so on. Their correlation is calculated in the usual way, as explained in Section 2.7.

The correlation r_{ME} has the usual interpretation in terms of the relationship between the two distance measures. Thus, r lies in the range from −1 to +1, with r = −1 indicating a perfect negative correlation, r = 0 indicating no correlation, and r = +1 indicating a perfect positive correlation. The significance or otherwise of the data will be the same for the test statistics Z and r, because in fact, there is a simple linear relationship between them.

Example 5.3: More on distances between samples of Egyptian skulls

Returning to the Egyptian skull data, we can ask the question of whether the distances given in Table 5.4, based on four skull measurements, are significantly related to the time differences between the five samples. This certainly does seem to be the case, but a definitive answer is provided by Mantel s test.

The sample times are approximately 4000 BC (early predynastic), 3300 BC (late predynastic), 1850 BC (12th and 13th Dynasties), 200 BC (Ptolemaic), and AD 150 (Roman). Comparing Penrose's distance measures with time differences (in thousands of years) therefore provides the lower diagonal distance matrices between the samples that are shown in Table 5.7. The correlation between the elements of these matrices is 0.954. It appears, therefore, that the distances agree very well.

There are 5! = 120 possible ways to reorder the five samples for one of the two matrices, and, consequently, there are 120 elements in the randomization distribution for the correlation. Of these, one is the observed correlation of 0.954, and another is a larger correlation. It follows that the observed correlation is significantly high at the (2/120)100% = 1.7% level, and there is evidence of a relationship between the two distance matrices. A one-sided test is appropriate, because there is no reason why the samples of skulls should become more similar as they get further apart in time.

Table 5.7 Penrose distances based on skull measurements and time differences (thousands of years) for five samples of Egyptian skulls

Penrose distances					Time distances				
—					—				
0.023	—				0.70	—			
0.216	0.163	—			2.15	1.45	—		
0.493	0.404	0.108	—		3.80	3.10	1.65	—	
0.736	0.583	0.244	0.066	—	4.15	3.45	2.00	0.35	—

The matrix correlation between Mahalanobis distances and time distance is 0.964. This is also significantly large at the 1.7% level when compared with the randomization distribution.

5.7   Computer programs

The calculation of distance and similarity measures is the first step in the analysis of multivariate data using cluster analysis and ordination methods. For this reason, the calculation of these measures is often easiest to do using computer programs that are designed for these methods. However, the clustering and ordination options of more general statistical packages can be used, or the R code described in the Appendix to this chapter. Computer programs for Mantel test on distance and similarity matrices are available from the website http://www.west-inc.com/computer-programs/, and these tests can also be carried out using the R code described in the Appendix to this chapter.

5.8   Discussion and further reading

The use of different measures of distance and similarity is the subject of continuing debates, indicating a lack of agreement about what is the best under different circumstances. The problem is that no measure is perfect, and the conclusions from an analysis may depend to some extent on which of several reasonable measures is used. The situation depends very much on the purpose of calculating the distances or similarities, and the nature of the data available.

The usefulness of the Mantel randomization method for testing for an association between two distance or similarity matrices has led to a number of proposals for methods to analyze relationships between three or more such matrices. These are reviewed by Manly (2007). At present, a major unresolved problem in this area relates to the question of how to take proper account of the effects of spatial correlation when, as is often the case, the items between which distances and similarities are measured tend to be similar when they are relatively close in space.

Peres-Neto and Jackson (2001) have suggested that the comparison between two distance matrices using a method called Procrustes analysis is better than the use of the Mantel test. Procrustes analysis was developed by Gower (1971) as a means of seeing how well two data configurations can be matched up after suitable manipulations. Peres-Neto and Jackson propose a randomization test to assess whether the matching that can be obtained with two distance matrices is significantly better than expected by chance.

Exercise

Consider the data in Table 1.3.

1. Standardize the environmental variables altitude, annual precipitation, annual maximum temperature, and annual minimum temperature to means of zero and standard deviations of one, and calculate Euclidean distances between all pairs of colonies using Equation 5.1 to obtain an environmental distance matrix.
2. Use the Pgi gene frequencies, converted to proportions, to calculate genetic distances between the colonies using Equation 5.5.
3. Carry out a Mantel matrix randomization test to determine whether there is a significant positive relationship between the environmental and genetic distances, and report your conclusions.
4. Explain why a significant positive relationship on a randomization test in a situation such as this could be the result of spatial correlations between the data for close colonies rather than from environmental effects on the genetic composition of colonies.

References

Gower, J.C. (1971). Statistical methods for comparing different multivariate analyses of the same data. In *Mathematics in the Archaeological and Historical Sciences* (eds F.R. Hodson, D.G. Kendall and P. Tautu), pp. 138–49. Edinburgh: Edinburgh University Press.

Gower, J.C. and Legendre, P. (1986). Metric and non-metric properties of dissimilarity coefficients. *Journal of Classification* 5: 5–48.

Higham, C.F.W., Kijngam, A. and Manly, B.F.J. (1980). Analysis of prehistoric canid remains from Thailand. *Journal of Archaeological Science* 7: 149–65.

Jackson, D.A., Somers, K.M. and Harvey, H.H. (1989). Similarity coefficients: Measures of co-occurrence and association or simply measures of co-occurrence. *American Naturalist* 133: 436–53.

Mahalanobis, P.C. (1948). Historic note on the D²-statistic. *Sankhya* 9: 237.

Manly, B.F.J. (2007). *Randomization, Bootstrap and Monte Carlo Methods in Biology.* 3rd Edn. London: Chapman and Hall.

Mantel, N. (1967). The detection of disease clustering and a generalized regression approach. *Cancer Research* 27: 209–20.

Mielke, P.W. (1978). Classification and appropriate inferences for Mantel and Varland's nonparametric multivariate analysis technique. *Biometrics* 34: 272–82.

Penrose, L.W. (1953). Distance, size and shape. *Annals of Eugenics* 18: 337–43.

Peres-Neto, P.R. and Jackson, D.A. (2001). How well do multivariate data sets match? The advantages of a Procrustean superimposition approach over a Mantel test. *Oecologia* 129: 169–78.

Appendix: Multivariate distance measures in R

A.1   Calculation of distance measures

The main function in R for distance calculations is dist(), from which the user can choose among six distance measures, the default being the Euclidean distance. The most usual arguments for dist() are either a numeric matrix or a data frame, and distances are computed between pairs of rows. If distances between columns are sought, the transpose operator t(), described in the Appendix for Chapter 2, must be applied. Additional options are available for formatting the output of the computed distance matrix, for example showing only the upper or lower triangular matrix.

The R script needed to obtain the Euclidean distances between dogs and related species (Example 5.1) is provided in the supplementary material at the book's website. According to the procedure seen in that example, first, the variables are standardized using scale() or with the function decostand() found in the package vegan (Oksanen et al., 2016), and then dist() is applied to the standardized data. In fact, vegan provides two other functions (vegdist and designdist) with further distance measures, some of them special for binary data, as described below.

For the calculation of the Penrose distance, no special R functions or packages are directly available. In contrast, a specific R function for computation of Mahalanobis distances is available, called D2.dist(). An R script for producing the Penrose and Mahalanobis distances between pairs of samples of Egyptian skulls, as in Table 5.4, can be found at the book's website. This script takes advantage of the immediate availability of the estimated pooled covariance matrix produced by function BoxM() from the package biotools, as described in the Appendix to Chapter 4, as a previous step in the calculation of Penrose and Mahalanobis distances.

Distances based on proportions are not available in R from the default packages. To implement in R the calculations of the distance index d_1 as given in Equation 5.5, the user may decide first to get the associated similarity index $s_1 = 1 - d_1$, as described in Section 5.4, as a measure of the niche overlap between two species. Ecologists name this index *Czekanowski's proportional similarity index*, and it can be accessed from packages rInSp, using function PSicalc() (Zaccarelli et al., 2013), or with EcoSimR, using function czeckanowski() (Gotelli et al., 2015). The distance d_2 (as defined by Equation 5.6) is known in ecology as *Pianka's niche overlap index*; it has been implemented in the spaa package (Zhang 2016), as the function niche.overlap(), assuming that the function is applied to the columns of a matrix community data. d_2 can also be computed with the function piankabio() located in the package pgirmess (Giraudoux 2016), while the similarity version of this index, $s_2 = 1 - d_2$ is present as the function pianka() in the package EcosimR().

In the case of multivariate binary distances in R, the user can make use of the options available in different packages, starting from the default package, `stats`, and some others, such as ade4 (Dray and Dufour, 2007) or vegan (Oksanen et al., 2016). Thus, the function `dist()` allows the computation of the Jaccard distance of the form $d_J = 1 - s_J$, where s_J is Jaccard's similarity index defined in Section 5.5:

```
dist(data, method="binary")
```

Here, `data` is a data object (matrix or data frame) containing either binary or nonnegative numbers, such that nonzero elements are converted into ones, and zero elements are kept intact. Jaccard's distances are also in vegan (Oksanen et al., 2016) as function vegdist(), and in ade4 (Dray and Dufour, 2007). In this latter package, ten different distances for binary data are available, all of them of the form $d = \sqrt{(1 - s)}$, where s is a similarity coefficient. Therefore, to compute a similarity index in ade4, like those shown in Section 5.5, the user must first invoke the corresponding distance d and then compute the similarity index as $s = 1 - d^2$.

A.2   The Mantel randomization test

The package vegan (Oksanen et al., 2016) offers the function `mantel` for the calculation of the Mantel statistic and its corresponding randomization test of significance. The simplest expression for invoking the mantel function is mantel(xdis, ydis).

The user can include the argument permutations= if the number of elements in the randomization distribution for the correlation is very large. This is not the case for Example 5.3, where the number of all possible permutations is only $5! = 120$.

The Mantel test is also available in the ade4 package (Dray and Dufour, 2007) via the function `mantel.rtest`. A script containing the functions mantel and mantel.test has been created and applied in the analysis of Egyptian skulls data (Example 5.3), and this script can be found at this book's website.

References

Dray, S. and Dufour, A.B. (2007). The ade4 package: Implementing the duality diagram for ecologists. *Journal of Statistical Software* 22(4): 1–20.

Giraudoux, P. (2016). pgirmess: Data analysis in ecology. R package version 1.6.4. https://CRAN.R-project.org/package=pgirmess

Gotelli, N.J., Hart, E.M. and Ellison, A.M. (2015). EcoSimR: Null model analysis for ecological data. R package version 0.1.0. http://github.com/gotellilab/EcoSimR

Oksanen, J., Blanchet, F.G., Friendly, M., Kindt, R., Legendre, P., McGlinn, D.,
    Minchin, P.R., et al. (2016). vegan: Community ecology package. R package
    version 2.4-0. http://CRAN.R-project.org/package=vegan
Zaccarelli, N., Mancinelli G. and Bolnick, D.I. (2013). RInSp: An R package for the
    analysis of individual specialisation in resource use. *Methods in Ecology and
    Evolution* 4(11): 1018–23.
Zhang, J. (2016). spaa: SPecies Association Analysis. R package version 0.2.2.
    https://CRAN.R-project.org/package=spaa

chapter six

Principal components analysis

6.1 Definition of principal components

The technique of principal components analysis was first described by Karl Pearson (1901). He apparently believed that this was the correct solution to some of the problems that were of interest to biometricians at that time, although he did not propose a practical method of calculation for more than two or three variables. A description of practical computing methods came much later from Hotelling (1933). Even then, the calculations were extremely daunting for more than a few variables, because they had to be done by hand. It was not until computers became generally available that the technique achieved widespread use.

Principal components analysis is one of the simplest of the multivariate methods. The object of the analysis is to take p variables $X_1, X_2, \ldots , X_p$, and find combinations of these to produce indices $Z_1, Z_2, \ldots , Z_p$ that are uncorrelated and in order of their importance in terms of the variation in the data. The lack of correlation means that the indices are measuring different dimensions of the data, and the ordering is such that $\mathrm{Var}(Z_1) \geq \mathrm{Var}(Z_2) \ldots \geq \mathrm{Var}(Z_p)$, where $\mathrm{Var}(Z_i)$ denotes the variance of Z_i. The Z indices are then the principal components. When doing a principal components analysis, there is always the hope that the variances of most of the indices will be so low as to be negligible. In that case, most of the variation in the full data set can be adequately described by the few Z variables with variances that are not negligible, and some degree of economy is then achieved.

Principal components analysis does not always work in the sense that a large number of original variables are reduced to a small number of transformed variables. Indeed, if the original variables are uncorrelated, then the analysis achieves nothing. The best results are obtained when the original variables are very highly correlated, positively or negatively. If that is the case, then it is quite conceivable that 20 or more original variables can be adequately represented by two or three principal components. If this desirable state of affairs does occur, then the important principal components will be of some interest as measures of the underlying dimensions in the data. It will also be of value to know that there is a good deal of redundancy in the original variables, with most of them measuring similar things.

Table 6.1 Correlations between the five body measurements of female sparrows
calculated from the data of Table 1.1

	X_1	X_2	X_3	X_4	X_5
X_1, total length	1.000	—			
X_2, alar extent	0.735	1.000	—		
X_3, length of beak and head	0.662	0.674	1.000	—	
X_4, length of humerus	0.645	0.769	0.763	1.000	—
X_5, length of keel of sternum	0.605	0.529	0.526	0.607	1.000

Note: Only the lower part of the table is shown, because the correlation between X_i and X_j is
the same as the correlation between X_j and X_i.

Before describing the calculations involved in a principal components analysis, it is of value to look briefly at the outcome of the analysis
when it is applied to the data in Table 1.1 on five body measurements of
49 female sparrows. Details of the analysis are given in Example 6.1. In
this case, the five measurements are quite highly correlated, as shown
in Table 6.1. This is, therefore, good material for the analysis in question. It turns out that the first principal component has a variance of 3.62,
whereas the other components all have variances that are much less than
this (0.53, 0.39, 0.30, and 0.16). This means that the first principal component is by far the most important of the five components for representing
the variation in the measurements of the 49 birds. The first component is
calculated to be

$$Z_1 = 0.45\,X_1 + 0.46\,X_2 + 0.45\,X_3 + 0.47\,X_4 + 0.40\,X_5$$

where X_1–X_5 denote the measurements in Table 1.1 in order, after they have
been standardized to have zero means and unit standard deviations.

Clearly, Z_1 is essentially just an average of the standardized body
measurements, and it can be thought of as a simple index of size. The
analysis given in Example 6.1 therefore leads to the conclusion that most
of the differences between the 49 birds are a matter of size rather than
shape.

6.2   Procedure for a principal components analysis

A principal components analysis starts with data on p variables for n individuals, as indicated in Table 6.2. The first principal component is then the
linear combination of the variables $X_1, X_2, \ldots, X_p$:

$$Z_1 = a_{11}\,X_1 + a_{12}\,X_2 + a_{1p}\,X_p$$

Table 6.2 The form of data for a
principal components analysis, with
variables X_1 to X_p and observations on n
cases

Case	X_1	X_2	...	X_p
1	x_{11}	x_{12}	...	x_{1p}
2	x_{21}	x_{22}	...	x_{2p}
n	x_{n1}	x_{n2}	...	x_{np}

that varies as much as possible for the individuals, subject to the condition
that

$$a_{11}^2 + a_{12}^2 + ... + a_{1p}^2 = 1$$

Thus, $Var(Z_1)$, the variance of Z_1, is as large as possible given this constraint on the constants a_{1j}. The constraint is introduced because if this is
not done, then $Var(Z_1)$ can be increased by simply increasing any one of
the a_{1j} values.

The second principal component

$$Z_2 = a_{21}X_1 + a_{22}X_2 + ... + a_{2p}X_p$$

is chosen so that $Var(Z_2)$ is as large as possible, subject to the constraint
that

$$a_{21}^2 + a_{22}^2 + ... + a_{2p}^2 = 1$$

and also to the condition that Z_1 and Z_2 have zero correlation for the data.
The third principal component

$$Z_3 = a_{31}X_1 + a_{32}X_2 + ... + a_{3p}X_p$$

is such that $Var(Z_3)$ is as large as possible, subject to the constraint that

$$a_{31}^2 + a_{32}^2 + ... + a_{3p}^2 = 1$$

and also that Z_3 is uncorrelated with both Z_1 and Z_2. Further principal
components are defined by continuing in the same way. If there are p
variables, then there will be up to p principal components.

To use the results of a principal components analysis, it is not necessary to know how the equations for the principal components are derived. However, it is useful to understand the nature of the equations themselves. In fact, a principal components analysis involves finding the eigenvalues of the sample covariance matrix.

The calculation of the sample covariance matrix has been described in Sections 2.6 and 2.7. The covariance matrix is symmetric and has the form

$$
C = \begin{bmatrix}
c_{11} & c_{12} & \cdots & c_{1p} \\
c_{21} & c_{22} & \cdots & c_{2p} \\
\cdot & \cdot & & \cdot \\
\cdot & \cdot & & \cdot \\
c_{p1} & c_{p2} & \cdots & c_{pp}
\end{bmatrix}
$$

where the diagonal element c_{ii} is the variance of X_i, and the off-diagonal terms $c_{ij} = c_{ji}$ are the covariance of variables X_i and X_j.

The variances of the principal components are the eigenvalues of the matrix C. There are p of these eigenvalues, some of which may be zero, but negative eigenvalues are not possible for a covariance matrix. Assuming that the eigenvalues are ordered as $\lambda_1 \geq \lambda_2 \geq \ldots \geq \lambda_p \geq 0$, then λ_i corresponds to the ith principal component

$$
Z_i = a_{i1}X_1 + a_{i2}X_2 + \ldots + a_{ip}X_p
$$

In particular, $\mathrm{Var}(Z_i) = \lambda_i$ and the constants $a_{i1}, a_{i2}, \ldots, a_{ip}$ are the elements of the corresponding eigenvector, scaled so that

$$
a_{i1}^2 + a_{i2}^2 + \ldots + a_{ip}^2 = 1
$$

An important property of the eigenvalues is that they add up to the sum of the diagonal elements (the trace) of the matrix C. That is,

$$
\lambda_1 + \lambda_2 + \ldots + \lambda_p = c_{11} + c_{22} + \ldots + c_{pp}
$$

As c_{ii} is the variance of X_i and λ_i is the variance of Z_i, this means that the sum of the variances of the principal components is equal to the sum of the variances of the original variables. Therefore, in a sense, the principal components account for all the variation in the original data.

To avoid one or two variables having an undue influence on the principal components, it is usual to code the variables X_1, X_2, ..., X_p to have means of zero and variances of one at the start of an analysis. The matrix **C** then takes the form

$$\mathbf{C} = \begin{bmatrix} 1 & c_{12} & \cdots & c_{1p} \\ c_{21} & 1 & \cdots & c_{2p} \\ \cdot & \cdot & & \cdot \\ \cdot & \cdot & & \cdot \\ c_{p1} & c_{p2} & \cdots & 1 \end{bmatrix}$$

where $c_{ij} = c_{ji}$ is the correlation between X_i and X_j. In other words, the principal components analysis is carried out on the correlation matrix. In that case, the sum of the diagonal terms, and hence the sum of the eigenvalues, is equal to p, the number of X variables.

The steps in a principal components analysis are

1. Start by coding the variables X_1, X_2, ..., X_p, to have zero means and unit variances. This is usual, but is omitted in some cases where it is thought that the importance of variables is reflected in their variances.
2. Calculate the covariance matrix **C**. This is a correlation matrix if Step 1 has been done.
3. Find the eigenvalues λ_1, λ_2, ..., λ_p and the corresponding eigenvectors $\mathbf{a}_1$, $\mathbf{a}_2$, ..., $\mathbf{a}_p$. The coefficients of the ith principal component are then the elements of $\mathbf{a}_i$, while λ_i is its variance.
4. Discard any components that only account for a small proportion of the variation in the data. For example, starting with 20 variables, it might be found that the first three components account for 90% of the total variance. On this basis, the other 17 components may reasonably be ignored.

Example 6.1: Body measurements of female sparrows

Some mention has already been made of what happens when a principal components analysis is carried out on the data on five body measurements of 49 female sparrows (Table 1.1). This example is now considered in more detail.

It is appropriate to begin with Step 1 of the four parts of the analysis that have just been described. Standardization of the measurements ensures that they all have equal weight in the analysis.

Table 6.3 The eigenvalues and eigenvectors of the correlation matrix for five measurements on 49 female sparrows

Component	Eigenvalue	Eigenvectors (coefficients for the principal components)				
		X_1	X_2	X_3	X_4	X_5
1	3.616	0.452	0.462	0.451	0.471	0.398
2	0.532	−0.051	0.300	0.325	0.185	−0.877
3	0.386	0.691	0.341	−0.455	−0.411	−0.179
4	0.302	−0.420	0.548	−0.606	0.388	0.069
5	0.165	0.374	−0.530	−0.343	0.652	−0.192

Note: The eigenvalues are the variances of the principal components. The eigenvectors give the coefficients of the standardized X variables used to calculate the principal components.

Omitting standardization would mean that the variables X_1 and X_2, which vary most over the 49 birds, would tend to dominate the principal components.

The covariance matrix for the standardized variables is the correlation matrix. This has already been given in lower triangular form in Table 6.1. The eigenvalues of this matrix are found to be 3.616, 0.532, 0.386, 0.302, and 0.164. These add to 5.000, the sum of the diagonal terms in the correlation matrix. The corresponding eigenvectors are shown in Table 6.3, standardized so that the sum of the squares of the coefficients is one for each of them. These eigenvectors then provide the coefficients of the principal components.

The eigenvalue for a principal component indicates the variance that it accounts for out of the total variances of 5.000. Thus, the first principal component accounts for $(3.616/5.000)100\% = 72.3\%$ of the total variance. Similarly, the other principal components in order account for 10.6%, 7.7%, 6.0%, and 3.3%, respectively, of the total variance. Clearly, the first component is far more important than any of the others.

Another way of looking at the relative importance of principal components is in terms of their variance in comparison with the variance of the original variables. After standardization, the original variables all have variances of 1.0. The first principal component, therefore, has a variance of 3.616 original variables. However, the second principal component has a variance of only 0.532 of that of one of the original variables, while the other principal components account for even less variation. This confirms the importance of the first principal component in comparison with the others.

The first principal component is

$$Z_1 = 0.452X_1 + 0.462X_2 + 0.451X_3 + 0.471X_4 + 0.398X_5$$

where X_1 to X_5 are the standardized variables. The coefficients of the X variables are nearly equal, and this is clearly an index of the size of the sparrows. It seems, therefore, that about 72.3% of the variation in the data is related to size differences among the sparrows.

The second principal component is

$$Z_2 = -0.051\,X_1 + 0.300\,X_2 + 0.325\,X_3 + 0.185\,X_4 - 0.877\,X_5$$

This is a contrast between variables X_2 (alar extent), X_3 (length of beak and head), and X_4 (length of humerus), on the one hand, and variable X_5 (length of the keel of the sternum), on the other. That is to say, Z_2 will be high if X_2, X_3, and X_4 are high but X_5 is low. On the other hand, Z_2 will be low if X_2, X_3, and X_4 are low but X_5 is high. Hence, Z_2 represents a shape difference between the sparrows. The low coefficient of X_1 (total length) means that the value of this variable does not affect Z_2 very much. The other principal components can be interpreted in a similar way. They therefore represent other aspects of shape differences.

The values of the principal components may be useful for further analyses. They are calculated in the obvious way from the standardized variables. Thus, for the first bird, the original variable values are $x_1 = 156$, $x_2 = 245$, $x_3 = 31.6$, $x_4 = 18.5$, and $x_5 = 20.5$. These standardize to $x_1 = (156 - 157.980)/3.654 = -0.542$, $x_2 = (245 - 241.327)/5.068 = 0.725$, $x_3 = (31.6 - 31.459)/0.795 = 0.177$, $x_4 = (18.5 - 18.469)/0.564 = 0.055$, and $x_5 = (20.5 - 20.827)/0.991 = -0.330$, where in each case the variable mean for the 49 birds has been subtracted, and a division has been made by the sample standard deviation for the 49 birds. The value of the first principal component for the first bird is therefore

$$Z_1 = 0.452 \times (-0.542) + 0.462 \times 0.725 + 0.451 \times 0.177 + 0.47\,1 \times 0.055$$

$$+\ 0.398 \times (-0.330)$$

$$= 0.064$$

The second principal component for the same bird is

$$Z_2 = -\,0.051 \times (-0.542) + 0.300 \times 0.725 + 0.325 \times 0.177 + 0.185 \times 0.055$$
$$-\,0.877 \times (-0.330)$$
$$= 0.602$$

The other principal components can be calculated in a similar way.

The birds being considered were picked up after a severe storm. The first 21 of them recovered, while the other 28 died. A question of some interest is, therefore, whether the survivors and nonsurvivors show any differences. It has been shown in Example 4.1 that there is no evidence of any differences in mean values. However, in Example 4.2, it has been shown that the survivors seem to have been less variable than the nonsurvivors. The situation will now be considered in terms of principal components.

Table 6.4 Comparison between survivors and nonsurvivors in terms of means and standard deviations of principal components

Principal component	Mean		Standard deviation	
	Survivors	Nonsurvivors	Survivors	Nonsurvivors
1	−0.100	0.075	1.506	2.176
2	0.004	−0.003	0.684	0.776
3	−0.140	0.105	0.522	0.677
4	0.073	−0.055	0.563	0.543
5	0.023	−0.017	0.411	0.408

The means and standard deviations of the five principal components are shown in Table 6.4 separately for survivors and non-survivors. None of the mean differences between survivors and nonsurvivors are significant from t-tests, and none of the standard deviation differences are significant on F-tests. However, Levene's test on deviations from medians (described in Chapter 4) just gives a significant difference between the variation of principal component 1 for survivors and nonsurvivors on a one-sided test at the 5% level. The assumption for the one-sided test is that, if anything, nonsurvivors were more variable than survivors. The variation is not significantly different for survivors and nonsurvivors with Levene's test on the other principal components. As principal component 1 measures overall size, it seems that stabilizing selection may have acted against very large and very small birds.

Figure 6.1 shows a plot of the values of the 49 birds for the first two principal components, which between them account for 82.9%

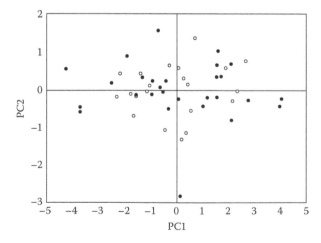

Figure 6.1 Plot of 49 female sparrows against values for the first two principal components, PC1 and PC2. ○ = survivor, ● = nonsurvivor.

of the variation in the data. The figure shows quite clearly how birds with extreme values for the first principal component failed to survive. Indeed, there is a suggestion that this was true for principal component 2 as well.

It is important to realize that some computer programs may give the principal components as shown with this example but with the signs of the coefficients of the body measurements reversed. For example, Z_2 might be shown as

$$Z_2 = 0.051X_1 - 0.300X_2 - 0.325X_3 - 0.185X_4 + 0.877X_5$$

This is not a mistake. The principal component is still measuring exactly the same aspect of the data, but in the opposite direction.

Example 6.2: Employment in European countries

As a second example of a principal components analysis, consider the data in Table 1.5 on the percentages of people employed in nine industry sectors in Europe. The correlation matrix for the nine variables is shown in Table 6.5. Overall, the values in this matrix are not particularly high, which indicates that several principal components will be required to account for the variation in the data.

The eigenvalues of the correlation matrix, with percentages of the total of 9.000 in parentheses, are 3.112 (34.6%), 1.809 (20.1%), 1.496 (16.6%), 1.063 (11.8%), 0.710 (7.9%), 0.311 (3.5%), 0.293 (3.3%), 0.204 (2.3%), and 0.000 (0.0%). The last eigenvalue is zero because the sum of the nine variables being analyzed is 100% before standardization. The principal component corresponding to this eigenvalue has the value zero for all the countries, and hence has a zero variance. If any linear

Table 6.5 The correlation matrix for percentages employed in nine industry groups in 30 countries in Europe in lower diagonal form, calculated from the data in Table 1.5

	AGR	MIN	MAN	PS	CON	SER	FIN	SPS	TC
AGR	1.000	—							
MIN	0.316	1.000	—						
MAN	-0.254	-0.672	1.000	—					
PS	-0.382	-0.387	0.388	1.000	—				
CON	-0.349	-0.129	-0.034	0.165	1.000	—			
SER	-0.605	-0.407	-0.033	0.155	0.473	1.000	—		
FIN	-0.176	-0.248	-0.274	0.094	-0.018	0.379	1.000	—	
SPS	-0.811	-0.316	0.050	0.238	0.072	0.388	0.166	1.000	—
TC	-0.487	0.045	0.243	0.105	-0.055	-0.085	-0.391	0.475	1.000

Note: The variables are the percentages employed in AGR, agriculture, forestry, and fishing; MIN, mining and quarrying; MAN, manufacturing; PS, power and water supplies; CON, construction; SER, services; FIN, finance; SPS, social and personal services; TC, transport and communications.

combination of the original variables in a principal components analysis is constant, then this must of necessity result in one of the eigenvalues being zero.

This example is not as straightforward as the previous one. The first principal component only accounts for about 35% of the variation in the data, and four components are needed to account for 83% of the variation. It is a matter of judgment as to how many components are important. It can be argued that only the first four should be considered, because these are the ones with eigenvalues greater than one. To some extent, the choice of the number of components that are important will depend on the use that is going to be made of them. For the present example, it will be assumed that a small number of indices are required to present the main aspects of differences between the countries, and for simplicity, only the first two components will be examined further. Between them, they account for about 55% of the variation in the original data.

The first component is

$$Z_1 = 0.51(AGR) + 0.37(MIN) - 0.25(MAN) - 0.31(PS) - 0.22(CON)$$

$$-0.38(SER) - 0.13(FIN) - 0.42(SPS) - 0.21(TC)$$

where the abbreviations for variables are defined in Table 6.5. As the analysis has been done on the correlation matrix, the variables in this equation are the original percentages after they have each been standardized to have a mean of zero and a standard deviation of one. From the coefficients of Z_1, it can be seen that it is a contrast between the numbers engaged in agriculture, forestry and fishing (AGR) and mining and quarrying (MIN), and the numbers engaged in other occupations.

The second component is

$$Z_2 = -0.02(AGR) + 0.00(MIN) + 0.43(MAN) + 0.11(PS) - 0.24(CON)$$
$$- 0.41(SER) - 0.55(FIN) + 0.05(SPS) + 0.52(TC)$$

which primarily contrasts the numbers in manufacturing (MAN) and transport and communications (TC) with the numbers in construction (CON), service industries (SER), and finance (FIN).

Figure 6.2 shows a plot of the 30 countries against their values for Z_1 and Z_2. The picture is certainly rather meaningful in terms of what is known about the countries. Most of the traditional Western democracies are grouped with slightly negative values for Z_1 and values for Z_2 between about plus and minus one. Gibraltar and Albania stand out as having rather distinct employment patterns, while the remaining countries lie in a band ranging from the former Yugoslavia ($Z_1 = -1.2$, $Z_2 = 2.2$) to Turkey ($Z_1 = 3.2$, $Z_2 = -0.3$).

As with the previous example, it is possible that some computer programs will produce the principal components shown here, but

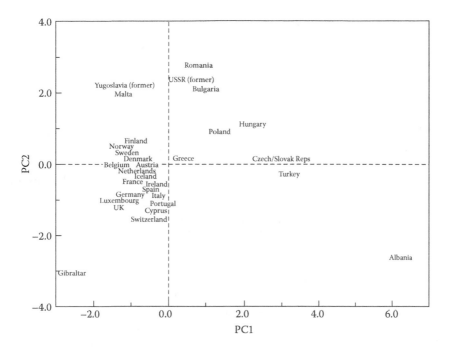

Figure 6.2 European countries plotted against the first two principal components for employment variables.

with the signs of the coefficients of the original variables reversed. The components still measure the same aspects of the data, but with the high and low values reversed.

6.3   Computer programs

The Appendix to this chapter provides the R code for carrying out a principal components analysis, and many standard statistical packages will carry out this analysis, because it is one of the most common types of multivariate analysis in use. When the analysis is not mentioned as an option in a package, it may still be possible to do the required calculations as a special type of factor analysis, as explained in Chapter 7. In that case, care will be needed to ensure that there is no confusion between the principal components and the factors, which are the principal components scaled to have unit variances.

This confusion can also occur with some programs that claim to be carrying out a principal component analysis. Instead of providing the values of the principal components (with variances equal to eigenvalues), they provide values of the principal components scaled to have variances of one.

6.4   Further reading

Principal components analysis is covered in almost all texts on multivariate analysis, and in greater detail by Jolliffe (1986) and Jackson (1991). Social scientists may also find the shorter monograph by Dunteman (1989) to be helpful.

Exercises

1. Table 6.6 shows six measurements on each of 25 pottery goblets excavated from prehistoric sites in Thailand, with Figure 6.3 illustrating the typical shape and the nature of the measurements. The main

Table 6.6 Measurements (in centimeters) taken on 25 prehistoric goblets from Thailand

Goblet	X_1	X_2	X_3	X_4	X_5	X_6
1	13	21	23	14	7	8
2	14	14	24	19	5	9
3	19	23	24	20	6	12
4	17	18	16	16	11	8
5	19	20	16	16	10	7
6	12	20	24	17	6	9
7	12	19	22	16	6	10
8	12	22	25	15	7	7
9	11	15	17	11	6	5
10	11	13	14	11	7	4
11	12	20	25	18	5	12
12	13	21	23	15	9	8
13	12	15	19	12	5	6
14	13	22	26	17	7	10
15	14	22	26	15	7	9
16	14	19	20	17	5	10
17	15	16	15	15	9	7
18	19	21	20	16	9	10
19	12	20	26	16	7	10
20	17	20	27	18	6	14
21	13	20	27	17	6	9
22	9	9	10	7	4	3
23	8	8	7	5	2	2
24	9	9	8	4	2	2
25	12	19	27	18	5	12

Note: The data were kindly provided by Professor C.F.W. Higham of the University of Otago, New Zealand. The variables are defined in Figure 6.3.

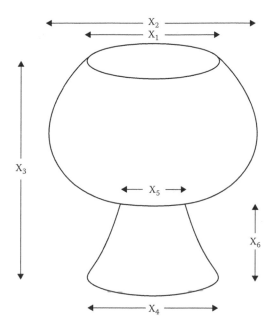

Figure 6.3 Measurements made on pottery goblets from Thailand.

question of interest for these data concerns similarities and differences between the goblets, with obvious questions being whether it
is possible to display the data graphically to show how the goblets
are related, and if so, whether there are any obvious groupings of
similar goblets and any goblets that are particularly unusual. Carry
out a principal components analysis and see whether the values of
the principal components help to answer these questions.

One point that needs consideration with this exercise is the extent
to which differences between goblets are due to shape differences
rather than size differences. It may well be considered that two goblets that are almost the same shape but have very different sizes are
really similar. The problem of separating size and shape differences
has generated a considerable scientific literature, which will not be
considered here. However, it can be noted that one way to remove
the effects of size involves dividing the measurements for a goblet
by the total height of the body of the goblet. Alternatively, the measurements on a goblet can be expressed as a proportion of the sum of
all measurements on that goblet. These types of standardization of
variables will ensure that the data values are similar for two goblets
with the same shape but different sizes.

2. Table 6.7 shows estimates of the average protein consumption from
   different food sources for the inhabitants of 25 European countries

Table 6.7 Protein consumption (grams per person per day) in 25 European countries

Country	Red meat	White meat	Eggs	Milk	Fish	Cereals	Starchy foods	Pulses, nuts, and oilseeds	Fruit and vegetables	Total
Albania	10	1	1	9	0.0	42	1	6	2	72
Austria	9	14	4	20	2.0	28	4	1	4	86
Belgium	14	9	4	18	5.0	27	6	2	4	89
Bulgaria	8	6	2	8	1.0	57	1	4	4	91
Czechoslovakia	10	11	3	13	2.0	34	5	1	4	83
Denmark	11	11	4	25	10.0	22	5	1	2	91
East Germany	8	12	4	11	5.0	25	7	1	4	77
Finland	10	5	3	34	6.0	26	5	1	1	91
France	18	10	3	20	6.0	28	5	2	7	99
Greece	10	3	3	18	6.0	42	2	8	7	99
Hungary	5	12	3	10	0.0	40	4	5	4	83
Ireland	14	10	5	26	2.0	24	6	2	3	92
Italy	9	5	3	14	3.0	37	2	4	7	84
The Netherlands	10	14	4	23	3.0	22	4	2	4	86
Norway	9	5	3	23	10.0	23	5	2	3	83
Poland	7	10	3	19	3.0	36	6	2	7	93
Portugal	6	4	1	5	14.0	27	6	5	8	76
Romania	6	6	2	11	1.0	50	3	5	3	87
Spain	7	3	3	9	7.0	29	6	6	7	77
Sweden	10	8	4	25	8.0	20	4	1	2	82
Switzerland	13	10	3	24	2.0	26	3	2	5	88
United Kingdom	17	6	5	21	4.0	24	5	3	3	88
USSR	9	5	2	17	3.0	44	6	3	3	92
West Germany	11	13	4	19	3.0	19	5	2	4	80
Yugoslavia	4	5	1	10	1.0	56	3	6	3	89

as published by Weber (1973). Use principal components analysis to investigate the relationships between the countries on the basis of these variables.

References

Dunteman, G.H. (1989). *Principal Components Analysis*, Newbury Park, CA: Sage.

Hotelling, H. (1933). Analysis of a complex of statistical variables into principal components. *Journal of Educational Psychology* 24: 417–41; 498–520.

Jackson, J.E. (1991). *A User's Guide to Principal Components*, New York: Wiley.

Jolliffe, I.T. (2002). *Principal Component Analysis*. 2nd Edn. New York: Springer.

Pearson, K. (1901). On lines and planes of closest fit to a system of points in space. *Philosophical Magazine* 2: 557–72.

Weber, A. (1973). *Agrarpolitik im Spannungsfeld der Internationalen Ernährungspolitik.* Kiel: Institut für Agrapolitik und Marktlehre.

Appendix: Principal Components Analysis (PCA) in R

The default R installation provides two computational methods for principal components analysis, performed by two functions, `princomp()` and `prcomp()`, which are loaded each time R is invoked. The former uses an algorithm that closely follows the procedure described in Section 6.2, based on the calculation of eigenvalues of the correlation matrix, or the covariance matrix if this is desired. In matrix algebra terminology, this technique is known as the *spectral decomposition of the covariance or correlation matrix*. On the other hand, `prcomp()` applies a method called the *singular value decomposition* (SVD) (Anton, 2013), a general procedure of matrix factorization, useful for handling matrices that are singular or nearly singular. The application of SVD to principal component analysis is supported by two facts: first, that the nonzero singular values of any real matrix **M** are the square roots of the nonzero eigenvalues of both **MM**^{T} and **M**^{T}**M**, the product of a matrix by its transpose, and vice versa; second, that the covariance or the correlation matrices can be expressed as the multiplication of these two matrices. In general, `princomp()` and `prcomp()` produce similar results. However, the implementation of the SVD algorithm is more accurate from a numerical point of view. Thus, `prcomp()` or any other R function for principal component analysis based on the SVD procedure is preferable.

R scripts with comments are provided in this book's website as computational aids for getting the results in Example 6.1, the analysis of the Bumpus sparrow data, and Example 6.2, the analysis of employment data in European countries. The basic command used in these examples is

```
pca.results<-prcomp(data, scale=TRUE,...)
```

The option `scale=TRUE` means that the principal components are computed on the correlation matrix. It is worth noticing that the object `pca.results` produced by `prcomp()` contains the singular values of the correlation or the covariance matrix. These values, labeled with the heading *Standard Deviation*, are revealed when the function `print(pca.results)` is executed. The eigenvalues are simply these standard deviations squared. In addition to the singular values, the eigenvectors, and the values for each principal component, two-dimensional graphical summaries of the PCA can be built by means of the function `plot()`, like the plots shown in Figures 6.1 and 6.2. It is also possible to produce a variation of the plot for the first two principal components through the command `biplot(pca.results)`. A biplot is a graphical summary of the PCA in which the first two principal components, plotted as points, are simultaneously displayed with a projection of the variables in the two-dimensional reduced space, plotted as arrows. See Gower and Hand (1996) for further details.

Several R packages offer functions for principal component analysis in addition to `princomp` and `prcompr`. A list of some of these packages and functions is given here. The one to use can be chosen based on the descriptions provided for each package in the corresponding help files or manuals.

Package	Function for PCA	Reference
`stats`	`princomp()`	R documentation (R Core Team 2016)
`stats`	`prcomp()`	R documentation (R Core Team 2016)
`FactoMineR`	`PCA()`	Le et al. (2008)
`ade4`	`dudi.pca()`	Dray and Dufour (2007)
`vegan`	`rda()`	Oksanen et al. (2016)
`amap`	`acp()`	Lucas (2014)

References

Anton, H. (2013). *Elementary Linear Algebra*. 11th Edn. New York: Wiley.

Dray, S. and Dufour, A.B. (2007). The ade4 package: Implementing the duality diagram for ecologists. *Journal of Statistical Software* 22(4): 1–20.

Gower, J.C. and Hand, D.J. (1996). *Biplots. Monographs on Statistics and Applied Probability*. London: Chapman & Hall.

Le, S., Josse, J. and Husson, F. (2008). FactoMineR: An R package for multivariate analysis. *Journal of Statistical Software* 25(1): 1–18.

Lucas, A. (2014). amap: Another Multidimensional Analysis Package. R package version 0.8-14. https://CRAN.R-project.org/package=amap

Oksanen, J., Blanchet, F.G., Friendly, M., Kindt, R., Legendre, P., McGlinn, D., Minchin, P.R., et al. (2016). vegan: Community Ecology Package. R package version 2.4-0. http://CRAN.R-project.org/package=vegan

R Core Team (2016). *R: A Language and Environment for Statistical Computing*. Vienna: R Foundation for Statistical Computing. https://www.r-project.org/

chapter seven

Factor analysis

7.1   *The factor analysis model*

Factor analysis has similar aims to principal components analysis. The basic idea is still that it may be possible to describe a set of p variables X_1, X_2, ..., X_p in terms of a smaller number of indices or factors, and in the process get a better understanding of the relationship between these variables. There is, however, one important difference. Principal components analysis is not based on any particular statistical model, whereas factor analysis is based on a model.

Charles Spearman is credited with the early development of factor analysis. He studied the correlations between students' test scores of various types and noted that many observed correlations could be accounted for by a simple model (Spearman, 1904). For example, in one case, he obtained the matrix of correlations shown in Table 7.1 for how boys in a preparatory school scored on tests in Classics, French, English, mathematics, discrimination of pitch, and music. He noted that this matrix has the interesting property that any two rows are almost proportional if the diagonals are ignored. Thus, for rows Classics and English, there are the ratios

$$\frac{0.83}{0.67} \approx \frac{0.70}{0.64} \approx \frac{0.66}{0.54} \approx \frac{0.63}{0.51} \approx 1.2$$

Based on this observation, Spearman suggested that the six test scores could be described by the equation

$$X_i = a_i F + e_i$$

where:
  X_i is the ith score after it has been standardized to have a mean of zero and a standard deviation of one for all the boys
  a_i is a constant
  F is a factor value, which has a mean of zero and a standard deviation of one for all the boys
  e_i is the part of X_i that is specific to the ith test only

Table 7.1 Correlations between test scores for boys in a preparatory school

	Classics	French	English	Mathematics	Discrimination of pitch	Music
Classics	1.00	0.83	0.78	0.70	0.66	0.63
French	0.83	1.00	0.67	0.67	0.65	0.57
English	0.78	0.67	1.00	0.64	0.54	0.51
Mathematics	0.70	0.67	0.64	1.00	0.45	0.51
Discrimination of pitch	0.66	0.65	0.54	0.45	1.00	0.40
Music	0.63	0.57	0.51	0.51	0.40	1.00

Source: Data from Spearman, C., *Am. J. Psychol.*, 15, 201–93, 1904.

Spearman showed that a constant ratio between the rows of a correlation matrix follows as a consequence of these assumptions, and therefore, this is a plausible model for the data.

Apart from the constant correlation ratios, it also follows that the variance of X_i is given by

$$
\begin{aligned}
\operatorname{Var}(X_i) &= \operatorname{Var}(a_i F + e_i) \\
&= \operatorname{Var}(a_i F) + \operatorname{Var}(e_i) \\
&= a_i^2 \operatorname{Var}(F) + \operatorname{Var}(e_i) \\
&= a_i^2 + \operatorname{Var}(e_i)
\end{aligned}
$$

because a_i is a constant, F and e_i are assumed to be independent, and the variance of F is assumed to be unity. Also, because $\operatorname{Var}(X_i) = 1$,

$$
1 = a_i^2 + \operatorname{Var}(e_i)
$$

Hence, the constant a_i, which is called the *factor loading*, is such that its square is the proportion of the variance of X_i that is accounted for by the factor.

On the basis of his work, Spearman formulated his two-factor theory of mental tests. According to this theory, each test result is made up of two parts: one that is common to all the tests (general intelligence) and another that is specific to the test in question. Later, this theory was modified to allow each test result to consist of a part due to several common factors plus a part specific to the test. This gives the general factor analysis model, which states that

$$
X_i = a_{i1} F_1 + a_{i2} F_2 + \ldots + a_{im} F_m + e_i
$$

where:

X_i	is the ith test score with mean zero and unit variance
a_{i1} to a_{im}	are the factor loadings for the ith test
F_1 to F_m	are m uncorrelated common factors, each with mean zero and unit variance
e_i	is specific only to the ith test, is uncorrelated with any of the common factors, and has zero mean

With this model,

$$
\operatorname{Var}(X_i) = 1 = a_{i1}^2 \operatorname{Var}(F_1) + a_{i2}^2 \operatorname{Var}(F_2) + \ldots + a_{im}^2 \operatorname{Var}(F_m) + \operatorname{Var}(e_i)
$$

$$
= a_{i1}^2 + a_{i2}^2 + \ldots + a_{im}^2 + \operatorname{Var}(e_i)
$$

where:

$a_{i1}^2 + a_{i2}^2 + \ldots + a_{im}^2$ is called the *communality* of X_i (the part of its variance that is related to the common factors)

$Var(e_i)$ is called the *specificity* of X_i (the part of its variance that is unrelated to the common factors)

It can also be shown that the correlation between X_i and X_j is

$$r_{ij} = a_{i1}a_{j1} + a_{i2}a_{j2} + \ldots + a_{im}a_{jm}$$

Hence, two test scores can only be highly correlated if they have high loadings on the same factors. Furthermore, $-1 \le a_{ij} \le +1$, as the communality cannot exceed one.

7.2   Procedure for a factor analysis

The data for a factor analysis have the same form as for a principal components analysis. That is, there are p variables with values for these n individuals, as shown in Table 6.2.

There are three stages to a factor analysis. To begin with, provisional factor loadings a_{ij} are determined. One approach starts with a principal components analysis and neglects the principal components after the first m, which are then taken to be the m factors. The factors found in this way are then uncorrelated with each other and are also uncorrelated with the specific factors. However, the specific factors are not uncorrelated with each other, which means that one of the assumptions of the factor analysis model does not hold. This may not matter much, providing that the communalities are high.

In whatever way the provisional factor loadings are determined, it is possible to show that they are not unique. If F_1, F_2, ..., F_m are the provisional factors, then linear combinations of these of the form

$$F^*_1 = d_{11}F_1 + d_{12}F_2 + \ldots + d_{1m}F_m$$

$$F^*_2 = d_{21}F_1 + d_{22}F_2 + \ldots + d_{2m}F_m$$

.

.

.

$$F^*_m = d_{m1}F_1 + d_{m2}F_2 + \ldots + d_{mm}F_m$$

can be constructed that are uncorrelated and explain the data just as well as the provisional factors. Indeed, there are an infinite number of alternative solutions for the factor analysis model. This leads to the second stage in the analysis, which is called *factor rotation*. At this stage, the provisional factors are transformed to find new factors that are easier to interpret. To rotate or to transform in this context means essentially to choose the d_{ij} values in the equations in the above equations.

The last stage of an analysis involves calculating the factor scores. These are the values of the rotated factors $F^*_1, F^*_2, \ldots, F^*_m$ for each of the n individuals for which data are available.

Generally, the number of factors (m) is up to the user, although it may sometimes be suggested by the nature of the data. When a principal components analysis is used to find a provisional solution, a rough rule of thumb involves choosing m to be the number of eigenvalues greater than unity for the correlation matrix of the test scores. The logic here is the same as was explained in the previous chapter on principal components analysis. A factor associated with an eigenvalue of less than unity accounts for less variation in the data than one of the original test scores. In general, increasing m will increase the communalities of variables. However, communalities are not changed by factor rotation.

Factor rotation can be orthogonal or oblique. With orthogonal rotation, the new factors are uncorrelated, like the provisional factors. With oblique rotation, the new factors are correlated. Whichever type of rotation is used, it is desirable that the factor loadings for the new factors should be either close to zero or very different from zero. A near-zero a_{ij} means that X_i is not strongly related to the factor F_j. A large positive or negative value of a_{ij} means that X_i is determined by F_j to a large extent. If each test score is strongly related to some factors but not related to the others, then this makes the factors easier to identify than would otherwise be the case.

One method of orthogonal factor rotation that is often used is called *varimax rotation*. This is based on the assumption that the interpretability of factor j can be measured by the variance of the squares of its factor loadings, that is, the variance of $a_{1j}^2, a_{2j}^2, \ldots, a_{pj}^2$. If this variance is large, then the a_{ij} values tend to be either close to zero or close to unity. Varimax rotation, therefore, maximizes the sum of these variances for all of the factors. Kaiser (1958) first suggested this approach. Later, he modified it slightly by normalizing the factor loadings before maximizing the variances of their squares, because this appears to give improved results. Varimax rotation can, therefore, be carried out with or without Kaiser normalization. Numerous other methods for orthogonal rotation have been proposed. However, varimax rotation seems to be a good standard approach.

Sometimes, factor analysts are prepared to give up the idea of the factors being uncorrelated so that the factor loadings should be as simple as possible. An oblique rotation may then give a better solution than an orthogonal one. Again, there are numerous methods available to do the oblique rotation.

A method for calculating the factor scores for individuals based on principal components is described in the section below. There are other methods available, so the one to be used will depend on the computer package or R code being used for an analysis.

7.3   Principal components factor analysis

It has been remarked above that one way to do a factor analysis is to begin with a principal components analysis and use the first few principal components as unrotated factors. This has the virtue of simplicity, although as the specific factors e_1, e_2, ... , e_p are correlated, the factor analysis model is not quite correct. Sometimes, factor analysts do a principal components factor analysis first and then try other approaches afterward.

The method for finding the unrotated factors is as follows. With p variables, there will be the same number of principal components. These are linear combinations of the original variables

$$Z_1 = b_{11}X_1 + b_{12}X_2 + ... + b_{1p}X_p$$

$$Z_2 = b_{21}X_1 + b_{22}X_2 + ... + b_{2p}X_p$$

$$.$$

$$.$$

$$.$$

$$Z_p = b_{p1}X_1 + b_{p2}X_2 + ... + b_{pp}X_p \tag{7.1}$$

where the b_{ij} values are given by the eigenvectors of the correlation matrix. This transformation from X values to Z values is orthogonal, so that the inverse relationship is simply

$$X_1 = b_{11}Z_1 + b_{21}Z_2 + ... + b_{p1}Z_p$$

$$X_2 = b_{12}Z_1 + b_{22}Z_2 + ... + b_{p2}Z_p$$

$$.$$

$$.$$

$$.$$

$$X_p = b_{1p}Z_1 + b_{2p}Z_2 + ... + b_{pp}Z_p$$

For a factor analysis, only m of the principal components are retained, so the last equations become

$$X_1 = b_{11}Z_1 + b_{21}Z_2 + \ldots + b_{m1}Z_m + e_1$$

$$X_2 = b_{12}Z_1 + b_{22}Z_2 + \ldots + b_{m2}Z_m + e_2$$

.

.

.

$$X_p = b_{1p}Z_1 + b_{2p}Z_2 + \ldots + b_{mp}Z_m + e_p$$

where e_i is a linear combination of the principal components Z_{m+1} to Z_p. All that needs to be done now is to scale the principal components Z_1, Z_2, $\ldots$, Z_m to have unit variances, as required for factors. To do this, Z_i must be divided by its standard deviation, which is $\sqrt{\lambda_i}$, the square root of the corresponding eigenvalue in the correlation matrix. The equations then become

$$X_1 = \sqrt{\lambda_1}b_{11}F_1 + \sqrt{\lambda_2}b_{21}F_2 + \ldots + \sqrt{\lambda_m}b_{m1}F_m + e_1$$

$$X_2 = \sqrt{\lambda_1}b_{12}F_1 + \sqrt{\lambda_2}b_{22}F_2 + \ldots + \sqrt{\lambda_m}b_{m2}F_m + e_2$$

.

.

.

$$X_p = \sqrt{\lambda_1}b_{1p}F_1 + \sqrt{\lambda_2}b_{2p}F_2 + \ldots + \sqrt{\lambda_m}b_{mp}F_m + e_p$$

where $F_i = Z_i/\sqrt{\lambda_i}$. The unrotated factor model is then

$$X_1 = a_{11}F_1 + a_{12}F_2 + \ldots + a_{1m}F_m + e_1$$

$$X_2 = a_{21}F_1 + a_{22}F_2 + \ldots + a_{2m}F_m + e_2$$

.

.

.

$$X_p = a_{p1}F_1 + a_{p2}F_2 + \ldots + a_{pm}F_m + e_p \tag{7.2}$$

where $a_{ij} = \sqrt{\lambda_j} b_{ji}$.

After a varimax or other type of rotation, a new solution has the form

$$X_1 = g_{11}F^*_1 + g_{12}F^*_2 + \ldots + g_{1m}F^*_m + e_1$$
$$X_2 = g_{21}F^*_1 + g_{22}F^*_2 + \ldots + g_{2m}F^*_m + e_2$$

.

.

.

$$X_p = g_{p1}F^*_1 + g_{p2}F^*_2 + \ldots + g_{pm}F^*_m + e_p \qquad (7.3)$$

where F^*_i represents the new ith factor.

The values of the ith unrotated factor are just the values of the ith principal component after these have been scaled to have a variance of one. The values of the rotated factors are more complicated to obtain, but it can be shown that these are given by the matrix equation

$$F^* = XG(G'G)^{-1} \qquad (7.4)$$

where:

 F*   is an n × m matrix containing the values for the m rotated factors in its columns, with one row for each of the n original rows of data

 X   is the n × p matrix of the original data for p variables and n observations, after coding the variables X_1 to X_p to have means of zero and variances of one

 G   is the p × m matrix of rotated factor loadings given by Equation 7.3

7.4   *Using a factor analysis program to do principal components analysis*

Because many computer programs for factor analysis allow the option of using principal components as initial factors, it is possible to use the programs to do principal components analysis. All that has to be done is to extract the same number of factors as variables and not do any rotation. The factor loadings will then be as given by Equation 7.2, with $m = p$ and $e_1 = e_2 = \ldots = e_p = 0$. The principal components are given by Equation 7.1, with $b_{ij} = a_{ji}/\sqrt{\lambda_i}$, where λ_i is the ith eigenvalue.

Example 7.1: Employment in European countries

In Example 6.2, a principal components analysis was carried out on the percentages of people employed in nine industry groups in 30 countries in Europe for the years 1989 to 1995 (Table 1.5). It is of some interest to continue the examination of these data using a factor analysis model.

The correlation matrix for the nine percentage variables is given in Table 6.5, and the eigenvalues and eigenvectors of this correlation matrix are shown in Table 7.2. There are four eigenvalues greater than unity, suggesting that four factors should be considered, which is what will be done here.

The eigenvectors in Table 7.2 give the coefficients of the X variables for Equation 7.1. These are changed into factor loadings for four factors using Equation 7.2, to give the model

$$X_1 = +0.90 \cdot F_1 - 0.03 \cdot F_2 - 0.34 \cdot F_3 + 0.02 \cdot F_4 + e_1 \, (0.93)$$

$$X_2 = +0.66 \cdot F_1 - 0.00 \cdot F_2 + 0.63 \cdot F_3 + 0.12 \cdot F_4 + e_2 \, (0.85)$$

$$X_3 = -0.43 \cdot F_1 + 0.58 \cdot F_2 - 0.61 \cdot F_3 + 0.06 \cdot F_4 + e_3 \, (0.91)$$

$$X_4 = -0.56 \cdot F_1 + 0.15 \cdot F_2 - 0.36 \cdot F_3 + 0.02 \cdot F_4 + e_4 \, (0.46)$$

$$X_5 = -0.39 \cdot F_1 - 0.33 \cdot F_2 + 0.09 \cdot F_3 + 0.81 \cdot F_4 + e_5 \, (0.92)$$

$$X_6 = -0.67 \cdot F_1 - 0.55 \cdot F_2 + 0.08 \cdot F_3 + 0.17 \cdot F_4 + e_6 \, (0.79)$$

$$X_7 = -0.23 \cdot F_1 - 0.74 \cdot F_2 - 0.12 \cdot F_3 - 0.50 \cdot F_4 + e_7 \, (0.87)$$

$$X_8 = -0.76 \cdot F_1 + 0.07 \cdot F_2 + 0.44 \cdot F_3 - 0.03 \cdot F_4 + e_8 \, (0.88)$$

$$X_9 = +0.36 \cdot F_1 + 0.69 \cdot F_2 + 0.50 \cdot F_3 - 0.04 \cdot F_4 + e_9 \, (0.87)$$

Here, the values in parentheses are the communalities. For example, the communality for variable X_1 is $(0.90)^2 + (-0.03)^2 + (-0.34)^2 + (0.02)^2 = 0.93$. The communalities are quite high for all variables except X_4 (power supplies). Most of the variance for the other eight variables is, therefore, accounted for by the four common factors.

Factor loadings that are 0.50 or more (ignoring the sign) are bold in these equations. These large and moderate loadings indicate how the variables are related to the factors. It can be seen that X_1 is almost entirely accounted for by factor 1 alone, X_2 is a mixture of factor 1 and factor 3, X_3 is accounted for by factor 1 and factor 2, and so on. An undesirable property of this choice of factors is that five of the nine X variables are related strongly to two of the factors. This suggests that a factor rotation may provide a simpler model for the data.

Table 7.2 Eigenvalues and eigenvectors for the European employment data of Table 1.5

					Eigenvectors				
Eigenvalue	X_1 AGR	X_2 MIN	X_3 MAN	X_4 PS	X_5 CON	X_6 SER	X_7 FIN	X_8 SPS	X_9 TC
3.111	0.512	0.375	−0.246	−0.315	−0.222	−0.382	−0.131	−0.428	−0.205
1.809	−0.024	−0.000	0.432	0.109	−0.242	−0.408	−0.553	0.055	0.516
1.495	−0.278	0.516	−0.503	−0.292	0.071	0.064	−0.096	0.360	0.413
1.063	0.016	0.113	0.058	0.023	0.783	0.169	−0.489	−0.317	−0.042
0.705	0.025	−0.345	0.231	−0.854	−0.064	0.269	−0.133	0.046	0.023
0.311	−0.045	0.203	−0.028	0.208	−0.503	0.674	−0.399	−0.167	−0.136
0.293	0.166	−0.212	−0.238	0.065	0.014	−0.165	−0.463	0.619	−0.492
0.203	0.539	−0.447	−0.431	0.157	0.030	0.203	−0.026	−0.045	0.504
0.000	−0.582	−0.419	−0.447	−0.030	−0.129	−0.245	−0.191	−0.410	−0.061

Note: The variables are the percentages employed in nine industry groups: AGR, agriculture, forestry, and fishing; MIN, mining and quarrying; MAN, manufacturing; PS, power and water supplies; CON, construction; SER, services; FIN, finance; SPS, social and personal services; TC, transport and communications.

A varimax rotation with Kaiser normalization was carried out. This produced the model

$$X_1 = +\textbf{0.85} \cdot F_1 + 0.10 \cdot F_2 + 0.27 \cdot F_3 - 0.36 \cdot F_4 + e_1$$

$$X_2 = +0.11 \cdot F_1 + 0.30 \cdot F_2 + \textbf{0.86} \cdot F_3 - 0.10 \cdot F_4 + e_2$$

$$X_3 = -0.03 \cdot F_1 + 0.32 \cdot F_2 - \textbf{0.89} \cdot F_3 - 0.09 \cdot F_4 + e_3$$

$$X_4 = -0.19 \cdot F_1 - 0.04 \cdot F_2 - \textbf{0.64} \cdot F_3 + 0.14 \cdot F_4 + e_4$$

$$X_5 = -0.02 \cdot F_1 + 0.08 \cdot F_2 - 0.04 \cdot F_3 + \textbf{0.95} \cdot F_4 + e_5$$

$$X_6 = -0.35 \cdot F_1 - 0.48 \cdot F_2 - 0.15 \cdot F_3 + \textbf{0.65} \cdot F_4 + e_6$$

$$X_7 = -0.08 \cdot F_1 - \textbf{0.93} \cdot F_2 + 0.00 \cdot F_3 - 0.01 \cdot F_4 + e_7$$

$$X_8 = -\textbf{0.91} \cdot F_1 - 0.17 \cdot F_2 - 0.12 \cdot F_3 - 0.04 \cdot F_4 + e_8$$

$$X_9 = -\textbf{0.73} \cdot F_1 + \textbf{0.57} \cdot F_2 - 0.03 \cdot F_3 - 0.14 \cdot F_4 + e_9$$

The communalities are unchanged, and the factors are still uncorrelated. However, this is a slightly better solution than the previous one, as only X_9 is appreciably dependent on more than one factor.

At this stage, it is usual to try to put labels on factors. In the present case, this is not too difficult, based on the highest loadings only.

Factor 1 has a high positive loading for X_1 (agriculture, forestry, and fishing) and high negative loadings for X_8 (social and personal services) and X_9 (transport and communications). Therefore, it measures the extent to which people are employed in agriculture rather than services and communications. It can therefore be labeled "rural industries rather than social service and communication."

Factor 2 has high negative loadings for X_7 (finance) and a fairly high coefficient for X_9 (transport and communications). This can therefore be labeled "lack of finance industries."

Factor 3 has a high positive loading for X_2 (mining and quarrying), a high negative loading for X_3 (manufacturing), and a moderately high negative loading for X_4 (power supplies). This can therefore be labeled "mining rather than manufacturing."

Finally, factor 4 has a high positive loading for X_5 (construction) and a moderately high positive loading for X_6 (service industries). Therefore, "construction and service industries" seems to be a fair label in this case.

The **G** matrix of Equations 7.3 and 7.4 is given by the factor loadings shown in the second model in this example. For example, $g_{11} = 0.85$ and $g_{12} = 0.10$, to two decimal places. Using these loadings and carrying out the matrix calculations shown in Equation 7.4 provides the

Table 7.3 Rotated factor scores for 30 European countries

Country	Factor 1	Factor 2	Factor 3	Factor 4
Belgium	−0.97	−0.56	−0.10	−0.48
Denmark	−0.89	−0.47	−0.03	−0.67
France	−0.56	−0.78	−0.15	−0.25
Germany	0.05	−0.57	−0.47	0.58
Greece	0.48	0.19	−0.23	0.02
Ireland	0.28	−0.60	−0.36	0.03
Italy	0.25	−0.13	0.17	1.00
Luxembourg	−0.46	−0.36	0.02	0.92
Netherlands	−1.36	−1.56	−0.03	−2.09
Portugal	0.66	−0.45	−0.37	0.64
Spain	0.23	−0.11	−0.09	0.93
United Kingdom	−0.50	−1.14	−0.35	−0.04
Austria	0.18	0.05	−0.71	0.56
Finland	−0.78	−0.20	−0.21	−0.52
Iceland	−0.18	−0.04	−0.06	0.46
Norway	−1.36	−0.17	0.20	−0.42
Sweden	−1.20	−0.52	0.04	−0.74
Switzerland	0.12	−0.67	0.01	0.65
Albania	3.16	−1.82	1.76	−1.78
Bulgaria	0.47	1.56	−0.57	−0.65
Czech/Slovak Republics	−0.26	1.45	3.12	0.44
Hungary	−1.05	1.70	2.82	−0.15
Poland	0.97	0.71	−0.37	−0.42
Romania	1.11	1.73	−1.69	−0.81
USSR (Former)	0.08	2.09	−0.11	0.14
Yugoslavia (Former)	0.13	1.48	−1.70	0.17
Cyprus	0.46	−0.32	0.03	1.08
Gibraltar	−0.05	−1.05	0.08	3.26
Malta	−1.17	0.49	−0.79	−1.31
Turkey	2.15	0.07	0.15	−0.56

Note: Factor 1 is "rural industries rather than social service industries and communication," factor 2 is "lack of finance industries," factor 3 is "mining rather than manufacturing," and factor 4 is "construction industries."

values for the factor scores for each of the 30 countries in the original data set. These factor scores are shown in Table 7.3.

From studying the factor scores, it can be seen that the values for factor 1 emphasize the importance of rural industries rather than services and communications in Albania and Turkey. The values for factor 2 indicate that Bulgaria, Hungary, Romania, and the former USSR had few people employed in finance, but the Netherlands and Albania had large numbers employed in this area. The values for factor 3 contrast Albania and the Czech/Slovak Republics, with high levels of mining rather than manufacturing, with Romania and Yugoslavia, where the reverse is true. Finally, the values for factor 4 contrast Gibraltar, with high numbers in construction and service industries, with the Netherlands and Albania, where this is far from being the case.

It would be possible and reasonable to continue the analysis of this set of data by trying models with fewer factors and different methods of factor extraction. However, sufficient has been said already to indicate the general approach, so the example will be left at this point.

It should be kept in mind by anyone attempting to reproduce this analysis that different statistical packages may give the eigenvectors shown in Table 7.2 except that all the coefficients have their signs reversed. A sign reversal may also occur through a factor rotation, so that the loadings for a rotated factor are the opposite of what is shown above. Sign reversals like this just reverse the interpretation of the factor concerned. For example, if the loadings for the rotated factor 1 were the opposite of those shown above, then it would be interpreted as social and personal services, and transport and communications rather than rural industries.

7.5 Options in analyses

Computer programs for factor analysis, including different R codes, may allow a number of different options for the analysis, which is likely to be rather confusing for the novice in this area. Typically, there might be four or five methods for the initial extraction of factors and about the same number of methods for rotating these factors (including no rotation). This gives in the order of 20 different types of factor analysis that can be carried out, with results that will differ to some extent at least.

There is also the question of the number of factors to extract. Some packages may make an automatic choice, which may or may not be acceptable. The possibility of trying different numbers of factors therefore increases the choices for an analysis even more.

On the whole, it is probably best to avoid using too many options when first practicing factor analysis. The use of principal components as initial factors with varimax rotation, as used in the example in this chapter, is a reasonable start with any set of data. The maximum likelihood method for extracting factors is a good approach in principle, and might also be tried if this is available.

7.6   The value of factor analysis

Factor analysis is something of an art, and it is certainly not as objective as many statistical methods. For this reason, some statisticians are skeptical about its value. For example, Chatfield and Collins (1986) list six problems with factor analysis and conclude that "factor analysis should not be used in most practical situations." Similarly, Seber (2004) notes as a result of simulation studies that even if the postulated factor model is correct, then the chance of recovering it using available methods is not high.

On the other hand, factor analysis is widely used to analyze data and, no doubt, will continue to be widely used in the future. The reason for this is that users find the results useful for gaining insight into the structure of multivariate data. Therefore, if it is thought of as a purely descriptive tool, with limitations that are understood, it must take its place as one of the important multivariate methods. What should be avoided is carrying out a factor analysis on a single small sample that cannot be replicated and then assuming that the factors obtained must represent underlying variables that exist in the real world.

7.7   Discussion and further reading

Factor analysis is discussed in many texts on multivariate analysis, although, as noted in the previous section, the topic is sometimes not presented enthusiastically (Chatfield and Collins, 1986; Seber, 2004). Recent texts are generally more positive. For example, Rencher (2002) discusses at length the validity of factor analysis and why it often fails to work. He notes that there are many sets of data for which factor analysis should not be used, but others for which the method is useful.

Factor analysis as discussed in this chapter is often referred to as *exploratory factor analysis*, because it starts with no assumptions about the number of factors that exist or the nature of these factors. In this respect, it differs from what is called *confirmatory factor analysis*, which requires the number of factors and the factor structure to be specified in advance. In this way, confirmatory factor analysis can be used to test theories about the structure of the data.

Confirmatory factor analysis is more complicated to carry out than exploratory factor analysis. The details are described by Bernstein (1988, chapter 7), and Tabachnick and Fidell (2013). Confirmatory factor analysis is a special case of structural equation modeling, which is covered in Chapter 14 of the latter book.

Exercise

Using Example 7.1 as a model, carry out a factor analysis of the data in Table 6.7 on protein consumption from 10 different food sources for the inhabitants of 25 European countries. Identify the important factors underlying the observed variables and examine the relationships between the countries with respect to these factors.

References

Bernstein, I.H. (1988). *Applied Multivariate Analysis*. Berlin: Springer.

Chatfield, C. and Collins, A.J. (1986). *Introduction to Multivariate Analysis*. London: Chapman and Hall.

Kaiser, H.F. (1958). The varimax criterion for analytic rotation in factor analysis. *Psychometrika* 23: 187–200.

Rencher, A.C. (2002). *Methods of Multivariate Statistics*. 2nd Edn. New York: Wiley.

Seber, G.A.F. (2004). *Multivariate Observations*. New York: Wiley.

Spearman, C. (1904). "General intelligence", objectively determined and measured. *American Journal of Psychology* 15: 201–93.

Tabachnick, B.G. and Fidell, L.S. (2013). *Using Multivariate Statistics*. 6th Edn. Boston, MA: Pearson.

Appendix: Factor Analysis in R

In its default package stats, R offers the function factanal() as a maximum likelihood (ML) method for extracting factors, a topic noted briefly in Section 7.5. Thus, ML factor analysis is also considered the default factor analysis in R. However, in Section 7.5, it was also emphasized that there are different approaches in factor analysis, each approach associated with a particular algorithm. Psychometric researchers have been the most interested in applying the range of existing algorithms for factor analysis. This explains why the R package psych, created and maintained by Revelle (2016a), has been considered as the main tool for psychometric applications of several multivariate methods, including factor analysis.

The psych package offers the fa function, from which the user may choose one of five methods of factor analysis (minimum residual, principal axis, weighted least squares, generalized least squares, and maximum likelihood factor analysis). Nevertheless, none of these options follows exactly the algorithm described in Section 7.3, whereby a principal components analysis (PCA) is used to produce initial factors, followed by a varimax rotation and the calculation of factor scores, which are also known as Bartlett scores, using Equation 7.4. It is not difficult to execute most of the steps of this PCA factor analysis with the set of R functions already considered in previous chapters (e.g., prcomp, matrix multiplication, and matrix inversion). The particular step completing this algorithm, varimax rotation with Kaiser normalization, can be performed with the varimax() function implemented in the stats package. However, this way of doing a PCA factor analysis can be avoided with principal(), another function in the psych package. Although this function is thought just to be doing a PCA, its output is organized in such a way that the component loadings are more suitable for a typical factor analysis, showing the best m factors. The developer of psych argues that the presence of principal() in his package as a choice for factor analysis, in addition to the algorithms executed by the fa() function, is because "psychologists typically use PCA in a manner similar to factor analysis and thus the principal function produces output that is perhaps more understandable than that produced by princomp in the stats package" (Revelle, 2016b). The command required to replicate the factor analysis described in Chapter 7 is then

    principal(data, nfactors = 4, rotate = "varimax").

At this book's website, the reader will find two R scripts written to carry out the factor analysis performed for Example 7.1. One script does the calculations in the fastest way via the principal() function. The second script makes use of functions prcomp() and varimax(). It has been written for instructive purposes, so that the reader can follow in detail the application of Equations 7.1 through 7.4 in this chapter.

References

Revelle, W. (2016a). *PSYCH: Procedures for Personality and Psychological Research.* Evanston, IL: Northwestern University. http://CRAN.R-project.org/package=psych. Version = 1.6.6.

Revelle, W. (2016b). *An Overview of the psych Package: Vignette of psych Procedures for Psychological, Psychometric, and Personality Research.* https://cran.fhcrc.org/web/packages/psych/

chapter eight

Discriminant function analysis

8.1 The problem of separating groups

The problem that is addressed with discriminant function analysis is the extent to which it is possible to separate two or more groups of individuals, given measurements for these individuals on several variables. For example, with the data in Table 1.1 on five body measurements of 21 surviving and 28 nonsurviving sparrows, it is interesting to consider whether it is possible to use the body measurements to separate survivors and nonsurvivors. Also, for the data shown in Table 1.2 on four dimensions of Egyptian skulls for samples from five time periods, it is reasonable to consider whether the measurements can be used to assign skulls to different time periods.

In the general case, there will be m random samples from different groups, with sizes $n_1, n_2, \ldots n_m$, and values will be available for p variables $X_1, X_2, \ldots, X_p$ for each sample member. Thus, the data for a discriminant function analysis takes the form shown in Table 8.1. The data for a discriminant function analysis do not need to be standardized to have zero means and unit variances prior to the start of the analysis, as is usual with principal components and factor analysis. This is because the outcome of a discriminant function analysis is not affected in any important way by the scaling of individual variables.

8.2 Discrimination using Mahalanobis distances

One approach to discrimination is based on Mahalanobis distances, as defined in Section 5.3. The mean vectors for the m samples can be regarded as estimates of the true mean vectors for the groups. The Mahalanobis distances from the individual cases to the group centers can then be calculated, and each individual can be allocated to the group to which it is closest. This may or may not be the group that the individual actually came from, so the percentage of correct allocations is an indication of how well groups can be separated using the available variables.

This procedure is more precisely defined as follows. Let

$$\bar{\mathbf{x}}_i' = (\bar{x}_{1i}, \bar{x}_{2i}, \ldots, \bar{x}_{pi})'$$

Table 8.1 The form of data for a discriminant function
analysis with m groups with possibly different sizes,
and p variables measured on each individual case

Case	X_1	X_2	...	X_p	Group
1	x_{111}	x_{112}	...	x_{11p}	1
2	x_{211}	x_{212}	...	x_{21p}	1
⋮	⋮	⋮	⋮	⋮	⋮
n_1	x_{n_111}	x_{n_112}	...	x_{n_11p}	1
1	x_{121}	x_{122}	...	x_{12p}	2
2	x_{221}	x_{222}	...	x_{22p}	2
⋮	⋮	⋮	⋮	⋮	⋮
n_2	x_{n_221}	x_{n_222}	...	x_{n_22p}	2
1	x_{1m1}	x_{1m2}	...	x_{1mp}	m
2	x_{2m1}	x_{2m2}	...	x_{2mp}	m
⋮	⋮	⋮	⋮	⋮	⋮
n_m	x_{n_mm1}	x_{n_mm2}	...	x_{n_mmp}	m

denote the vector of mean values for the sample from the ith group, let $\mathbf{C}_i$
denote the covariance matrix for the same sample, and let $\mathbf{C}$ denote the
pooled sample covariance matrix, where these vectors and matrices are calculated as explained in Section 2.7. Then, the Mahalanobis distance from
an observation $\mathbf{x}' = (x_1, x_2, ..., x_p)'$ to the center of group i is estimated to be

$$\mathbf{D}_i^2 = (\mathbf{x} - \bar{\mathbf{x}}_i)'\mathbf{C}^{-1}(\mathbf{x} - \bar{\mathbf{x}}_i)$$

$$\sum_{r=1}^{p} \sum_{s=1}^{p} (x_r - \bar{x}_{ri})c^{rs}(x_s - \bar{x}_{si}) \tag{8.1}$$

where c^{rs} is the element in the rth row and the sth column of $\mathbf{C}^{-1}$. The
observation $\mathbf{x}$ is then allocated to the group for which $\mathbf{D}_i^2$ has the smallest
value.

8.3   *Canonical discriminant functions*

It is sometimes useful to be able to determine functions of the variables
$X_1, X_2, ..., X_p$ that in some sense separate the m groups as much as is possible. The simplest approach then involves taking a linear combination of
the X variables

$$Z = a_1X_1 + a_2X_2 + ... + a_pX_p$$

Table 8.2 An analysis of variance on the Z index

Source of variation	Degrees of freedom	Mean square	F-ratio
Between groups	m − 1	M_B	M_B/M_W
Within groups	N − m	M_W	
	N − 1		

for this purpose. Groups can be well separated using values of Z if the mean value of this variable changes considerably from group to group, with the values within a group being fairly constant.

One way to determine the coefficients $a_1, a_2, ..., a_p$ in the index involves choosing these so as to maximize the F-ratio for a one-way analysis of variance. Thus, if there are a total of N individuals in all the groups, an analysis of variance on Z values takes the form shown in Table 8.2. Hence, a suitable function for separating the groups can be defined as the linear combination for which the F-ratio M_B/M_W is as large as possible, as first suggested by Fisher (1936).

When this approach is used, it turns out that it may be possible to determine several linear combinations for separating groups. In general, the number available, e.g., s, is the smaller of p and m − 1. The linear combinations are referred to as canonical discriminant functions.

The first function

$$Z_1 = a_{11}X_1 + a_{12}X_2 + ... + a_{1p}X_p$$

gives the maximum possible F-ratio for a one-way analysis of variance for the variation within and between groups. If there is more than one function, then the second one

$$Z_2 = a_{21}X_1 + a_{22}X_2 + ... + a_{2p}X_p$$

gives the maximum possible F-ratio on a one-way analysis of variance, subject to the condition that there is no correlation between Z_1 and Z_2 within groups. Further functions are defined in the same way. Thus, the ith canonical discriminant function

$$Z_i = = a_{i1}X_1 + a_{i2}X_2 + ... + a_{ip}X_p$$

is the linear combination for which the F-ratio on an analysis of variance is maximized, subject to Z_i being uncorrelated with $Z_1, Z_2, ...$, and Z_{i-1} within groups.

Finding the coefficients of the canonical discriminant functions turns out to be an eigenvalue problem. The within-sample matrix of sums of squares and cross products, **W**, and the total sample matrix of sums of squares and cross products matrix, **T**, are calculated as described in Section 4.7. From these, the between-groups matrix

$$\mathbf{B} = \mathbf{T} - \mathbf{W}$$

can be determined. Next, the eigenvalues and eigenvectors of the matrix **W**⁻¹**B** have to be found. If the eigenvalues are $\lambda_1 > \lambda_2 > ... > \lambda_s$, then λ_i is the ratio of the between-group sum of squares to the within-group sum of squares for the ith linear combination, Z_i, while the elements of the corresponding eigenvector $\mathbf{a}_i' = (a_{i1}, a_{i2}, ..., a_{ip})$ are the coefficients of the X variables for this index.

The canonical discriminant functions $Z_1, Z_2, ..., Z_s$ are linear combinations of the original variables chosen in such a way that Z_1 reflects group differences as much as possible, Z_2 captures as much as possible of the group differences not displayed by Z_1, Z_3 captures as much as possible of the group differences not displayed by Z_1 and Z_2, and so on. The hope is that the first few functions are sufficient to account for almost all the important group differences. In particular, if only the first one or two functions are needed for this purpose, then a simple graphical representation of the relationship between the various groups is possible by plotting the values of these functions for the sample individuals.

8.4   Tests of significance

Several tests of significance are useful in conjunction with a discriminant function analysis. In particular, the T²-test of Section 4.3 can be used to test for a significant difference between the mean values for any pair of groups, while one of the tests described in Section 4.7 can be used to test for overall significant differences between the means for the m groups.

In addition, a test is sometimes used for testing whether the mean of the discriminant function Z_j differs significantly from group to group. This is based on the individual eigenvalues of the matrix **W**⁻¹**B**. For example, sometimes, the statistic

$$\varphi_j^2 = \{N - 1 - (p + m)/2\} \log_e (1 + \lambda_j)$$

is used, where N is the total number of observations in all groups. This statistic is then tested against the chi-squared distribution with p + m − 2j degrees of freedom (df), and a significantly large value is considered to provide evidence that the population mean values of Z_j vary from group to group. Alternatively, the sum $\varphi_j^2 + \varphi_{j+1}^2 + \ldots + \varphi_s^2$ is sometimes used for testing for group differences related to discriminant functions Z_j to Z_s. This is tested against the chi-squared distribution, with the df being the sum of those associated with the component terms. Other tests of a similar nature are also used.

Unfortunately, these tests are suspect to some extent, because the jth discriminant function in the population may not appear as the jth discriminant function in the sample due to sampling errors. For example, the estimated first discriminant function (corresponding to the largest eigenvalue for the sample matrix $\mathbf{W}^{-1}\mathbf{B}$) may in reality correspond to the second discriminant function for the population being sampled. Simulations indicate that this can upset the chi-squared tests described in the previous paragraph quite seriously. Therefore, it seems that the tests should not be relied on to decide how many of the obtained discriminant functions represent real group differences. See Harris (2001) for an extended discussion of the difficulties surrounding these tests and alternative ways to examine the nature of group differences.

One useful type of test that is valid, at least for large samples, involves calculating the Mahalanobis distance from each of the observations to the mean vector for the group containing the observation, as discussed in Section 5.3. These distances should follow approximately chi-squared distributions with p df. Hence, if an observation is very significantly far from the center of its group in comparison with the chi-squared distribution, then this brings into question whether the observation really came from the group in question.

8.5   Assumptions

The methods discussed so far in this chapter are based on two assumptions. First, for all the methods, the population within-group covariance matrix should be the same for all groups. Second, for tests of significance, the data should be multivariate normally distributed within groups.

In general, it seems that multivariate analyses that assume normality may be upset quite badly if this assumption is not correct. This contrasts with the situation with univariate analyses such as regression and analysis of variance, which are generally quite robust to this assumption. However, a failure of one or both assumptions does not necessarily mean that a discriminant function analysis is a waste of time. For example, it may well turn out that excellent discrimination is possible on data from nonnormal distributions, although it may not then be simple to establish the statistical significance of the group differences. Furthermore, discrimination methods that do not require the equality of population covariance matrices are available, as discussed in Section 8.12.

Example 8.1: Comparison of samples of Egyptian skulls

This example concerns the comparison of the values for four measurements on male Egyptian skulls for five samples ranging in age from the early predynastic period (circa 4000 BC) to the Roman period (circa AD 150). The data are shown in Table 1.2, and it has already been established that the mean values differ significantly from sample to sample (Example 4.3), with the differences tending to increase with the time difference between samples (Example 5.3).

The within-sample and total sample matrices of sums of squares and cross products are calculated as described in Section 4.7. They are found to be

$$
\mathbf{W} = \begin{bmatrix}
3061.67 & 5.33 & 11.47 & 291.30 \\
5.33 & 3405.27 & 754.00 & 412.53 \\
11.47 & 754.00 & 3505.97 & 164.33 \\
291.30 & 412.53 & 164.33 & 1472.13
\end{bmatrix}
$$

and

$$
\mathbf{T} = \begin{bmatrix}
3563.89 & -222.81 & -615.16 & 426.73 \\
-222.81 & 3635.17 & 1046.28 & 346.47 \\
-615.16 & 1046.28 & 4309.27 & -16.40 \\
426.73 & 346.47 & -16.40 & 1533.33
\end{bmatrix}
$$

The between-sample matrix is therefore

$$
\mathbf{B} = \mathbf{T} - \mathbf{W} = \begin{bmatrix}
502.83 & -228.15 & -626.63 & 135.43 \\
-228.15 & 229.91 & 292.28 & -66.07 \\
-626.63 & 292.28 & 803.30 & -180.73 \\
135.43 & -66.07 & -180.73 & 61.30
\end{bmatrix}
$$

The eigenvalues of $\mathbf{W}^{-1}\mathbf{B}$ are found to be $\lambda_1 = 0.437$, $\lambda_2 = 0.035$, $\lambda_3 = 0.015$, and $\lambda_4 = 0.002$, and the corresponding canonical discriminant functions are

$$Z_1 = -0.0107X_1 + 0.0040X_2 + 0.0119X_3 - 0.0068X_4$$

$$Z_2 = 0.0031X_1 + 0.0168X_2 - 0.0046X3 - 0.0022X_4$$

$$Z_3 = -0.0068X_1 + 0.0010X_2 + 0.0000X_3 + 0.0247X_4 \qquad (8.2)$$

and

$$Z_4 = 0.0126X_1 - 0.0001X_2 + 0.0112X_3 + 0.0054X_4$$

Because λ_1 is much larger than the other eigenvalues, it is apparent that most of the sample differences are described by Z_1 alone.

The X variables in Equation 8.2 are the values as shown in Table 1.2 without standardization. The nature of the variables is illustrated in Figure 1.1, from which it can be seen that large values of Z_1 correspond to skulls that are tall but narrow, with long jaws and short nasal heights.

The Z_1 values for individual skulls are calculated in the obvious way. For example, the first skull in the early predynastic sample has $X_1 = 131$ mm, $X_2 = 138$ mm, $X_3 = 89$ mm, and $X_4 = 49$ mm. Therefore, for this skull

$$Z_1 = -0.0107 \times 131 + 0.0040 \times 138 + 0.0119 \times 89 - 0.0068 \times 49 = -0.124$$

The means and standard deviations found for the Z_1 values for the five samples are shown in Table 8.3. It can be seen that the mean of Z_1 has become lower over time, indicating a trend toward shorter, broader skulls with short jaws but relatively large nasal heights. This is, however, very much an average change. If the 150 skulls are allocated to the samples to which they are closest according to the Mahalanobis distance function of Equation 8.1, then only 51 of them (34%) are allocated to the samples to which they really belong

Table 8.3 Means and standard deviations for the discriminant function Z_1 with five samples of Egyptian skulls

Sample	Mean	Standard deviation
Early predynastic	−0.029	0.097
Late predynastic	−0.043	0.071
12th and 13th Dynasties	−0.099	0.075
Ptolemaic	−0.143	0.080
Roman	−0.167	0.095

Table 8.4 Results obtained when 150 Egyptian skulls are allocated to the group for which they have the minimum Mahalanobis distance

Source group	Number allocated to group					Total
	1	2	3	4	5	
1	12	8	4	4	2	30
2	10	8	5	4	3	30
3	4	4	15	2	5	30
4	3	3	7	5	12	30
5	2	4	4	9	11	30

(Table 8.4). Thus, although this discriminant function analysis has been successful in pinpointing the changes in skull dimensions over time, it has not produced a satisfactory method for aging individual skulls.

Example 8.2: Discriminating between groups of European countries

The data shown in Table 1.5 on the percentages employed in nine industry groups in 30 European countries have already been examined by principal components analysis and by factor analysis (Examples 6.2 and 7.1). Here, they will be considered from the point of view of the extent to which it is possible to discriminate between groups of countries on the basis of employment patterns. In particular, four natural groups existed in the period when the data were collected. These were (1) the European Union (EU) countries of Belgium, Denmark, France, Germany, Greece, Ireland, Italy, Luxembourg, the Netherlands, Portugal, Spain, and the United Kingdom; (2) the European Free Trade Area (EFTA) countries of Austria, Finland, Iceland, Norway, Sweden, and Switzerland; (3) the Eastern European countries of Albania, Bulgaria, the Czech/Slovak Republics, Hungary, Poland, Romania, the former USSR, and the former Yugoslavia; and (4) the other countries of Cyprus, Gibraltar, Malta, and Turkey. These four groups can be used as a basis for a discriminant function analysis. Wilks' lambda test (Section 4.7) gives a very highly significant result ($p < 0.001$), so there is very clear evidence, overall, that these groups are meaningful.

Apart from rounding errors, the percentages in the nine industry groups add to 100% for each of the 30 countries. This means that any one of the nine percentage variables can be expressed as 100 minus the remaining variables. It is, therefore, necessary to omit one of the variables from the analysis to carry out the analysis. The last variable, the percentage employed in transport and communications, has therefore been omitted for the analysis that will now be described.

The number of canonical variables is three in this example, this being the minimum of the number of variables ($p = 8$) and the number of groups minus one ($m - 1 = 3$). These canonical variables are found to be

$$Z_1 = 0.427 \text{ AGR} + 0.295 \text{ MIN} + 0.359 \text{ MAN} + 0.339 \text{ PS} + 0.222 \text{ CON} +$$
$$0.688 \text{ SER} + 0.464 \text{ FIN} + 0.514 \text{ SPS}$$

$$Z_2 = 0.674 \text{ AGR} + 0.579 \text{ MIN} + 0.550 \text{ MAN} + 1.576 \text{ PS} + 0.682 \text{ CON} +$$
$$0.658 \text{ SER} + 0.349 \text{ FIN} + 0.682 \text{ SPS}$$

and

$$Z_3 = 0.732 \text{ AGR} + 0.889 \text{ MIN} + 0.873 \text{ MAN} + 0.410 \text{ PS} + 0.524 \text{ CON} +$$

$$0.895 \text{ SER} + 0.714 \text{ FIN} + 0.764 \text{ SPS}$$

Different computer programs are likely to output these canonical variables with all the signs reversed for the coefficients of one or more of the variables. Also, it may be desirable to reverse the signs that are output. Indeed, with this example, the output from the computer program had negative coefficients for all the variables with Z_1 and Z_2. The signs were therefore all reversed to make the coefficients positive. It is important to note that it is the original percentages employed that are to be used in these equations, rather than these percentages after they have been standardized to have zero means and unit variances.

The eigenvalues of $\mathbf{W^{-1}B}$ corresponding to the three canonical variables are $\lambda_1 = 5.349$, $\lambda_2 = 0.570$, and $\lambda_3 = 0.202$. The first canonical variable is therefore clearly the most important.

Because all the coefficients are positive for all three canonical variables, it is difficult to interpret what exactly they mean in terms of the original variables. It is helpful in this respect to consider instead the correlations between the original variables and the canonical variables, as shown in Table 8.5. This table includes the original variable TC (transport and communications), because the correlations for this variable are easily calculated once the values of Z_1 to Z_3 are known for all the European countries.

It can be seen that the first canonical variable has correlations above 0.5 for SER (services), FIN (finance), and SPS (social and personal services), and a correlation of -0.5 or less for AGR (agriculture, forestry, and fisheries), and MIN (mining). This canonical variable therefore represents service types of industry rather than traditional industries. There are no really large positive or negative correlations

Table 8.5 Correlations between the
original percentages in different
employment groups and the three
canonical variates

Group	Z_1	Z_2	Z_3
AGR	−0.50	0.37	0.09
MIN	−0.62	0.03	0.20
MAN	−0.02	−0.20	0.12
PS	0.17	0.18	−0.23
CON	0.14	0.26	−0.34
SER	0.82	−0.01	0.08
FIN	0.61	−0.36	−0.09
SPS	0.56	−0.19	−0.28
TC	−0.22	−0.47	−0.41

between the second canonical variate and the original variables. However, considering the largest correlations, it seems to represent agriculture and construction, with an absence of transport, communications, and financial services. Finally, the third canonical variable also shows no large correlations, but represents, if anything, an absence of transport, communication, and construction.

Plots of the countries against their values for the canonical variables are shown in Figure 8.1. The plot of the second canonical variable against the first one shows a clear distinction between the eastern countries on the left-hand side and the other groups on the left. There is no clear separation between the EU and EFTA countries, with Malta and Cyprus being in the same cluster. Turkey and Gibraltar from the "other" group of countries appear in the top at the right-hand side. It can be clearly seen how most separation occurs with the horizontal values for the first canonical variate. Based on the interpretation of the canonical variables given in the previous paragraph, it appears that in the eastern countries, there is an emphasis on traditional industries rather than service industries, whereas the opposite tends to be true for the other countries. Similarly, Turkey and Gibraltar stand out because of the emphasis on agriculture and construction rather than transport, communications, and financial services. For Gibraltar, there are apparently none engaged in agriculture, but a very high percentage in construction.

The plot of the third canonical variable against the first one shows no real vertical separation of the EU, EFTA, and Other groups of countries, although there are some obvious patterns, such as the Scandinavian countries appearing close together.

The discriminant function analysis has been successful in this example in separating the eastern countries from the others, with less success in separating the other groups. The separation is perhaps clearer than what was obtained using principal components, as shown in Figure 6.2.

8.6   Allowing for prior probabilities of group membership

Computer programs often allow many options for a discriminant function analysis. One situation is that the probability of membership is inherently different for different groups. For example, if there are two groups, it might be known that most individuals fall into group 1, while very few fall into group 2. In that case, if an individual is to be allocated to a group, it makes sense to bias the allocation procedure in favor of group 1. Thus, the process of allocating an individual to the group from which it has the smallest Mahalanobis distance should be modified. To allow for this, some computer programs permit prior probabilities of group membership to be taken into account in the analysis.

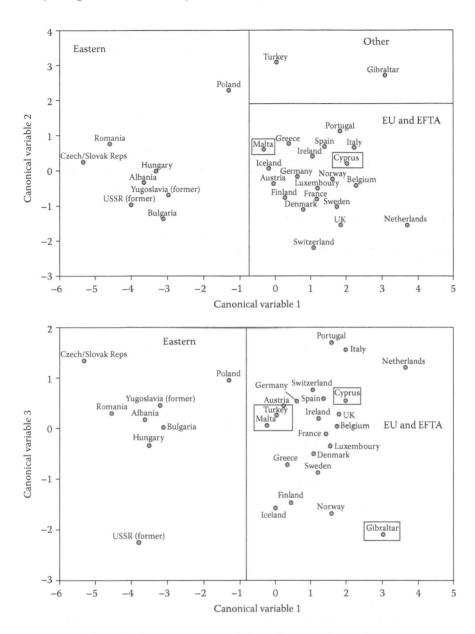

Figure 8.1 Plot of 30 European countries against their values for the first three canonical discriminant functions. Small boxes indicate countries in the other category that are not separated from the EU and EFTA groups.

8.7   Stepwise discriminant function analysis

Another possible modification of the basic analysis involves carrying it out in a stepwise manner. In this case, variables are added to the discriminant functions one by one until it is found that adding extra variables does not give significantly better discrimination. There are many different criteria that can be used for deciding which variables to include in the analysis and which to miss out.

A problem with stepwise discriminant function analysis is the bias that the procedure introduces into significance tests. Given enough variables, it is almost certain that some combination of them will produce significant discriminant functions by chance alone. If a stepwise analysis is carried out, then it is advisable to check its validity by rerunning it several times with a random allocation of individuals to groups to see how significant are the results obtained. For example, with the Egyptian skull data, the 150 skulls could be allocated completely at random to five groups of 30, the allocation being made a number of times, and a discriminant function analysis run on each random set of data. Some idea could then be gained of the probability of getting significant results through chance alone.

This type of randomization analysis to verify a discriminant function analysis is unnecessary for a standard nonstepwise analysis provided there is no reason to suspect the assumptions behind the analysis. It could, however, be informative in cases where the data are clearly not normally distributed within groups, or where the within-group covariance matrix is not the same for each group. For example, Manly (2007, Example 12.4) shows a situation in which the results of a standard discriminant function analysis are clearly suspect by comparison with the results of a randomization analysis.

8.8   Jackknife classification of individuals

Using an allocation matrix such as that shown in Table 8.4 must tend to have a bias in favor of allocating individuals to the group that they really come from. After all, the group means are determined from the observations in that group. It is, therefore, not surprising to find that an observation is closest to the center of the group where the observation helped to determine that center.

To overcome this bias, some computer programs carry out what is called a *jackknife classification* of observations. This involves allocating each individual to its closest group without using that individual to help determine a group center. In this way, any bias in the allocation is avoided. In practice, there is often not a great deal of difference between the straightforward classification and the jackknife classification, with the jackknife classification usually giving a slightly smaller number of correct allocations.

8.9 Assigning ungrouped individuals to groups

Some computer programs allow the input of data values for a number of individuals for which the true group is not known. It is then possible to assign these individuals to the group that they are closest to, in the Mahalanobis distance sense, on the assumption that they have to come from one of the m groups that are sampled. Obviously, in these cases, it will not be known whether the assignment is correct. However, the error in the allocation of individuals from known groups is an indication of how accurate the assignment process is likely to be. For example, the results shown in Table 8.4 indicate that allocating Egyptian skulls to different time periods using skull dimensions is liable to result in many errors.

8.10 Logistic regression

A rather different approach to discrimination between two groups involves making use of logistic regression. To explain how this is done, the more usual use of logistic regression will first be briefly reviewed.

The general framework for logistic regression is that there are m groups to be compared, with group i consisting of n_i items, of which Y_i exhibit a positive response (a success), and $n_i - Y_i$ exhibit a negative response (a failure). The assumption is then made that the probability of a success for an item in group i is given by

$$\pi_i = \frac{\exp\left(\beta_0 + \beta_1 x_{i1} + \beta_2 x_{i2} + \ldots + \beta_p x_{ip}\right)}{1 + \exp\left(\beta_0 + \beta_1 x_{i1} + \beta_2 x_{i2} + \ldots + \beta_p x_{ip}\right)} \tag{8.3}$$

where x_{ij} is the value of some variable X_j that is the same for all items in the group. In this way, the variables X_1 to X_p are allowed to influence the probability of a success, which is assumed to be the same for all items in the group, irrespective of the successes or failures of the other items in that or any other group. Similarly, the probability of a failure is $1 - \pi_i$ for all items in the ith group. It is permissible for some or all of the groups to contain just one item. Indeed, some computer programs only allow for this to be the case.

There need be no concern about arbitrarily choosing what to call a success and what to call a failure. It is easy to show that reversing these designations in the data simply results in all the β values and their estimates changing sign, and consequently changing π_i into $1 - \pi_i$.

The function that is used to relate the probability of a success to the X variables is called a *logistic function*. Unlike the standard multiple regression function, the logistic function forces estimated probabilities to lie within the range zero to one. It is for this reason that logistic regression is more sensible than linear regression as a means of modeling probabilities.

There are numerous computer programs available for fitting Equation 8.3 to data, that is, for estimating the values of β_0 to β_p, including R codes, as discussed in the Appendix to this chapter.

In the context of discrimination with two samples, three different types of situations have to be considered:

1. The data consist of a single random sample taken from a population of items that is itself divided into two parts. The application of logistic regression is then straightforward, and the fitted Equation 8.3 can be used to give an estimate of the probability of an item being in one part of the population (i.e., being a success) as a function of the values that the item possesses for variables X_1 to X_p. In addition, the distribution of success probabilities for the sampled items is an estimate of the distribution of these probabilities for the full population.
2. Separate sampling is used, whereby a random sample of size n_1 is taken from the population of items of one type (the successes), and an independent random sample of size n_2 is taken from the population of items of the second type (the failures). Logistic regression can still be used. However, the estimated probability of a success obtained from the estimated function must be interpreted in terms of the sampling scheme and the sample sizes used.
3. Groups of items are chosen to have particular values for the variables X_1 to X_p such that these variable values change from group to group. The number of successes in each group is then observed. In this case, the estimated logistic regression equation gives the probability of a success for an item conditional on the values that the item possesses for X_1 to X_p. The estimated function is, therefore, the same as for Situation (1), but the sample distribution of probabilities of a success is in no way an estimate of the distribution that would be found in the combined population of items that are successes or failures.

The following examples illustrate the differences between Situations (1) and (2), which are the ones that most commonly occur. Situation (3) is really just a standard logistic regression situation, and will not be considered further here.

Example 8.3: Storm survival of female sparrows (reconsidered)

The data in Table 1.1 consist of values for five morphological variables for 49 female sparrows taken in a moribund condition to Hermon Bumpus' laboratory at Brown University, Rhode Island, after a severe storm in 1898. The first 21 birds recovered and the remaining 28 died, and there is some interest in knowing whether it is possible to discriminate between these two groups on the basis of the five measurements. It has already been shown that there are no significant differences between the mean values of the variables for survivors and nonsurvivors (Example 4.1), although the nonsurvivors may have been more variable (Example 4.2). A principal components analysis has also confirmed the test results (Example 6.1).

This is a situation of Type (1) if the assumption is made that the sampled birds were randomly selected from the population of female sparrows in some area close to Bumpus' laboratory. Actually, the assumption of random sampling is questionable, because it is not clear exactly how the birds were collected. Nevertheless, the assumption will be made for this example.

The logistic regression option in many standard computer packages can be used to fit the model

$$\pi_i = \frac{\exp(\beta_0 + \beta_1 x_{i1} + \beta_2 x_{i2} + \ldots + \beta_5 x_{i5})}{1 + \exp(\beta_0 + \beta_1 x_{i1} + \beta_2 x_{i2} + \ldots + \beta_5 x_{i5})}$$

where:
X_1 = total length (mm)
X_2 = alar extent (mm)
X_3 = length of beak and head (mm)
X_4 = length of the humerus (mm)
X_5 = length of the sternum (mm)
π_i denotes the probability of the ith bird recovering from the storm

A chi-squared test for whether the variables account significantly for the difference between survivors and nonsurvivors gives the value 2.85 with five df, which is not at all significantly large when compared with chi-squared tables. There is, therefore, no evidence from this analysis that the survival status was related to the morphological variables. Estimated values for β_0 to β_5 are shown in Table 8.6,

Table 8.6 Estimates of the constant term and the coefficients of X variables when a logistic regression model is fitted to data on the survival of 49 female sparrows

Variable	β Estimate	Standard error	Chi-squared	p-Value
Constant	13.582	15.865	—	—
Total length	−0.163	0.140	1.36	0.244
Alar extent	−0.028	0.106	0.07	0.794
Length beak and head	−0.084	0.629	0.02	0.894
Length humerus	1.062	1.023	1.08	0.299
Length keel of sternum	0.072	0.417	0.03	0.864

Note: The chi-squared value is (estimated β value/standard error)². The p-value is the probability of a value this large from the chi-squared distribution with one degree of freedom. A small p-value (say, less than 0.05) provides evidence that the true value of the β parameter concerned is not equal to zero.

together with estimated standard errors and chi-squared statistics for testing whether the individual estimates differ significantly from zero. Again, there is no evidence of any significant effects.

Example 8.4: Comparison of two samples of Egyptian skulls

As an example of separate sampling, in which the sample size in the two groups being compared is not necessarily related in any way to the respective population sizes, consider the comparison between the first and last samples of Egyptian skulls for which data are provided in Table 1.2. The first sample consists of 30 male skulls from burials in the area of Thebes during the early predynastic period (circa 4000 BC) in Egypt, and the last sample consists of 30 male skulls from burials in the same area during the Roman period (circa AD 150). For each skull, measurements are available for X_1 = maximum breadth, X_2 = basibregmatic height, X_3 = basialveolar length, and X_4 = nasal height, all in millimeters (Figure 1.1). For the purpose of this example, it will be assumed that the two samples were effectively randomly chosen from their respective populations, although there is no way of knowing how realistic this is.

Obviously, the equal sample sizes in no way indicate that the population sizes in the two periods were equal. The sizes are, in fact, completely arbitrary, because many more skulls have been measured from both periods, and an unknown number of skulls have either not survived intact or not been found. Therefore, if the two samples are lumped together and treated as a sample of size 60 for the estimation of a logistic regression equation, then it is clear that the estimated probability of a skull with certain dimensions being from the early predynastic period may not really be estimating the true probability at all.

In fact, it is difficult to define precisely what is meant by the true probability in this example, because the population is not at all clear. A working definition is that the probability of a skull with specified dimensions being from the predynastic period is equal to the proportion of all skulls with the given dimensions that are from the predynastic period in a hypothetical population of all male skulls from either the predynastic or the Roman period that might have been recovered by archaeologists in the Thebes region.

It can be shown (Seber, 2004, p. 312) that if a logistic regression is carried out on a lumped sample to estimate Equation 8.3, then the modified equation

$$\pi_i = \frac{\exp\left(\beta_0 - \log_e\left\{(n_1P_2)/(n_2P_1)\right\} + \beta_1x_{i1} + \beta_2x_{i2} + \ldots + \beta_px_{ip}\right)}{1 + \exp\left(\beta_0 - \log_e\left\{(n_1P_2)/(n_2P_1)\right\} + \beta_1x_{i1} + \beta_2x_{i2} + \ldots + \beta_px_{ip}\right)} \quad (8.4)$$

is what really gives the probability that an item with the specified X values is a success. Here, Equation 8.4 differs from Equation 8.3 because of the term $\log_e\{(n_1P_2)/(n_2P_1)\}$ in the numerator and the

denominator, where P_1 is the proportion of items in the full population of successes and failures that are successes, and $P_2 = 1 - P_1$ is
the proportion of the population that are failures. This, then, means
that estimating the probability of an item with the specified X values
being a success requires that P_1 and P_2 are either known or can some-
how be estimated separately from the sample data to adjust the estimated logistic regression equation for the fact that the sample sizes n_1
and n_2 are not proportional to the population frequencies of successes
and failures. In the example being considered, this requires that estimates of the relative frequencies of predynastic and Roman skulls in
the Thebes area must be known to be able to estimate the probability
that a skull is predynastic, given the values that it possesses for the
variables X_1 to X_4.

Logistic regression was applied to the lumped data from the 60
predynastic and Roman skulls, with a predynastic skull being treated
as a success. The resulting chi-squared test for the extent to which
success is related to the X variables is 27.13 with four df. This is significantly large at the 0.1% level, giving very strong evidence of a relationship. The estimates of the constant term and the coefficients of
the X variables are shown in Table 8.7. It can be seen that the estimate
of β_1 is significantly different from zero at about the 1% level, and β_3
is significantly different from zero at the 2% level. Hence, X_1 and X_3
appear to be the important variables for discriminating between the
two types of skull.

The fitted function can be used to discriminate between the
two groups of skulls by assigning values for P_1 and $P_2 = 1 - P_1$ in
Equation 8.4. As already noted, it is desirable that these should correspond to the population proportions of predynastic and Roman
skulls. However, this is not possible, because these proportions are
not known. In practice, therefore, arbitrary values must be assigned.
For example, suppose that P_1 and P_2 are both set equal to 0.5. Then,
$\log_e\{(n_1 P_2)/(n_2 P_1)\} = \log_e(1) = 0$, because $n_1 = n_2$, and Equations 8.3 and
8.4 become identical. The logistic function, therefore, estimates the

Table 8.7 Estimates of the constant term and the coefficients of X variables
when a logistic regression model is fitted to data on 30 predynastic and
30 Roman period male Egyptian skulls

Variable	β Estimate	Standard error	Chi-squared	p-Value
Constant	−6.732	13.081	—	—
Maximum breadth	−0.202	0.075	7.13	0.008
Basibregmatic height	0.129	0.079	2.66	0.103
Basialveolar length	0.177	0.073	5.84	0.016
Nasal height	−0.008	0.104	0.01	0.939

Note: See Table 8.6 for an explanation of the columns.

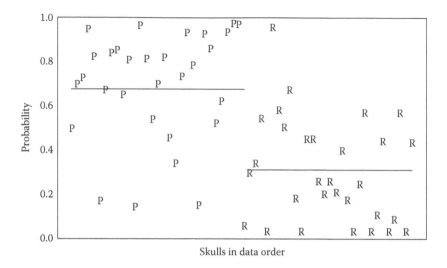

Figure 8.2 Values from the fitted logistic regression function plotted for 30 pre-dynastic (P) and 30 Roman (R) skulls. The horizontal lines indicate the average group probabilities.

probability of a skull being predynastic in a population with equal frequencies of predynastic and Roman skulls.

The extent to which the logistic equation is effective for discrimination is indicated in Figure 8.2, which shows the estimated values of π_i for the 60 sample skulls. There is a distinct difference in the distributions of values for the two samples, with the mean for predynastic skulls being about 0.7 and the mean for Roman skulls being about 0.3. However, there is also a considerable overlap between the distributions. As a result, if the sample skulls are classified as being predynastic when the logistic equation gives a value greater than 0.5 or as Roman when the equation gives a value of less than 0.5, then six predynastic skulls are misclassified as being Roman, and seven Roman skulls are misclassified as being predynastic.

8.11   *Computer programs*

Major statistical packages, including R (see the Appendix to this chapter), generally have a discriminant function option that applies the methods described in Sections 8.2 through 8.5, based on the assumption of normally distributed data. Because the details of the order of calculations, the way the output is given, and the terminology vary considerably, manuals may have to be studied carefully to determine precisely what is done by any individual program. Logistic regression is also fairly widely available. In some programs, there is the restriction that all items are assumed to have different values for X variables. However, it is more common for groups of items with common X values to be permitted.

8.12   Discussion and further reading

The assumption that samples are from multivariate distributions with the same covariance matrix, which is required for the use of the methods described in Sections 8.2 through 8.5, can be relaxed. If the samples being compared are assumed to come from multivariate normal distributions with different covariance matrices, then a method called *quadratic discriminant function analysis* can be applied. This option is also available in many computer packages. See Seber (2004, p. 297) for more information about this method and a discussion of its performance relative to the more standard linear discriminant function analysis.

Discrimination using logistic regression has been described in Section 8.10 in terms of the comparison of two groups. More detailed treatments of this method are provided by Hosmer et al. (2013) and Collett (2002). The method can also be generalized for discrimination between more than two groups if necessary, under several names, including *polytomous regression*. See Hosmer et al. (2013, Chapter 8) for more details. This type of analysis is a standard option in many computer packages.

Exercises

1. Consider the data in Table 4.5 for nine mandible measurements on samples from five different canine groups. Carry out a discriminant function analysis to see how well it is possible to separate the groups using the measurements.

2. Still considering the data in Table 4.5, investigate each canine group separately to see whether logistic regression shows a significant difference between males and females for the measurements. Note that in view of the small sample sizes available for each group, it is unreasonable to expect to fit a logistic function involving all nine variables with good estimates of parameters. Therefore, consideration should be given to fitting functions using only a subset of the variables.

References

Collett, D. (2002). *Modelling Binary Data*. 2nd Edn. Boca Raton, FL: Chapman and Hall/CRC.

Fisher, R.A. (1936). The utilization of multiple measurements in taxonomic problems. *Annals of Eugenics* 7: 179–88.

Harris, R.J. (2001). *A Primer on Multivariate Statistics*. 2nd Edn. New York: Psychology.

Hosmer, D.W., Lemeshow, S., and Sturdivant, R.X. (2013). *Applied Logistic Regression.* 3rd Edn. New York: Wiley.

Manly, B.F.J. (2007). *Randomization, Bootstrap and Monte Carlo Methods in Biology.* 3rd Edn. Boca Raton, FL: Chapman and Hall/CRC.

Seber, G.A.F. (2004). *Multivariate Observations.* New York: Wiley-Interscience.

Appendix: Discriminant Function Analysis in R

A.1 Canonical discriminant analysis in R

From a computational point of view, the method of canonical discriminant functions encompasses the eigenvalue analysis of $\mathbf{W}^{-1}\mathbf{B}$ (with $\mathbf{W}$ and $\mathbf{B}$ defined in Section 8.3), a task that can be carried out with the R-functions `eigen()` (described in the Appendix for Chapter 2) and `manova` (described in the Appendix for Chapter 4). This strategy is illustrated for Example 8.1 (the comparison of samples of Egyptian skulls) with an R script that can be downloaded from this book's website.

An alternative to using R programming is provided by the function `lda()` (linear discriminant analysis) included in the package MASS (Venables and Ripley, 2002). This uses two different methods of variable specification: either

$$\texttt{discan.object <- lda(group.f \sim X1 + X2 + ...,...)}$$

or

$$\texttt{discan.object <- lda(Xmat, group.f, ...)}$$

In the first method, `group.f` is a grouping factor and the variables X1, X2, ... are the discriminator variables, reminding us that discriminant analysis involves continuous independent variables and a categorical dependent variable (i.e., the `group.f` label). The second option assumes that Xmat is a matrix or data frame whose columns are the discriminators. It is possible to specify probabilities of group membership in `lda()` with the `prior` option.

It is important to notice that the canonical coefficients produced by `eigen()` and `lda()` differ from those shown in Equation 8.2, because R scales eigenvectors in several ways. For example, `eigen()` forces each eigenvector to have a unit norm. These variations are of no concern, as one set of coefficients can be computed from another set with a suitable linear transformation. In the case of `lda`, the canonical coefficients are normalized so that the within-groups covariance matrix $\mathbf{W}$ is spherical (i.e., $\mathbf{W}$ is a multiple of the identity matrix $\mathbf{I}$). Actually, the `print` method of `lda()` does not produce eigenvalues. Instead, it produces singular values, which are the ratio of the between- and within-group standard deviations of the linear discriminant variables. In addition, the output of `lda()` includes the proportion of trace, which is the proportion accounted by each eigenvalue of $\mathbf{W}^{-1}\mathbf{B}$ with respect to the sum of all the eigenvalues.

This proportion can be interpreted equivalently as the proportion of the between-group variance present on each discriminant axis.

When lda is executed, the user may choose to produce a classification table similar to Table 8.4 in Example 8.1 through the function predict, using

$$\text{table}\big(\text{group.f, predict}(\text{discan.object})\$\text{class}\big)$$

The function predict() accepts an additional parameter with the name of a data frame containing new data to be assigned to a particular group based on the canonical discriminant functions. It is worth noting that predict() is not necessary when considering jackknifing classification of individuals, which is a procedure that can be produced in lda() with the option CV=TRUE. Here, CV stands for *cross validation*, which is another name given in multivariate analysis for the jackknife classification method. An example of this command is

```
discan.object.cv <-lda (group. f ~ X1 + X2 +..., CV = TRUE,...)
```

Here, the content of the object discan.object.cv (a list) is different from that generated by lda() with CV=FALSE (the default). Now, discan. object.cv includes the vector class, and it is not difficult to build a table with original membership of individuals and their jackknife classification based on the discriminant analysis. See an R code exemplifying this procedure in the book's website.

The function lda() also includes a plot method, which is useful for displaying one, two, or more linear discriminant functions using

```
plot(discan.object, dimen,...)
```

The resulting plot depends on the parameter dimen, the number of dimensions chosen. See the R documentation for further details.

An alternative and helpful R function for plot visualization in canonical discriminant analysis is offered by candisc(), which is present in the package with the same name (Friendly and Fox, 2016). The function candisc uses a multivariate linear model like that produced by the lm() function. It generates its own scaling of eigenvectors, and its corresponding plot method permits the display of centroids or means of discriminant scores for each group. R codes containing lda and candisc functions are available in the book's website, as computational aids for the discriminant analyses described in Examples 8.1 and 8.2.

A.2   Discriminant analysis based on logistic regression in R

Logistic regression is available in R as one option of the function glm() for fitting generalized linear models (Hilbe, 2009). A typical logistic regression analysis is written as

```
model.logistic <-glm
```

$$\left(\text{Y} \sim \text{X1} + \text{X2} + ..., \text{family} = \text{binomial}(\text{link} = \text{"logit"}), ...\right)$$

where:
         Y   is a binary response variable
         X1, X2, ...   are explanatory variables

To use logistic regression for discriminant analysis of two samples, it is only necessary to code one sample as 1 and the other as 0 and assign these binary values to a new variable Y. The parameter estimates of the logistic regression can be obtained with the summary() function

$$\text{summary}(\text{model.logistic})$$

In addition, chi-squared tests for model comparison are accessible through the command anova(). Thus,

$$\text{anova}(\text{model.1, model.2, test} = \text{"Chisq"})$$

evaluates whether the fit of model.2 improves over the fit of model.1, assuming that the variables in model.1 is a subset of the variables in model.2. The chi-squared tests indicated in the Examples for Section 8.10 can be carried out in that way, using R: model.1 is the intercept-only model (i.e., no variables) and model.2 includes all the discriminators of interest. R codes are provided in the book's website exemplifying the use of glm and anova in Examples 8.3 and 8.4.

The extensions of linear discriminant analysis that are described in Sections 8.7 and 8.12 are available for the R user. The function stepclass located in the package klaR (Weihs et al., 2005) is what R offers for those interested in stepwise discriminant function analysis, while the function qda() in the package MASS (Venables and Ripley, 2002) allows the separation of groups using quadratic discriminant analysis. Finally, polytomous regression, the extension of two-group logistic regression for more than two groups, is accessible through the function multinom() from the package nnet (Venables and Ripley, 2002).

References

Friendly, M. and Fox, J. (2016). candisc: Visualizing Generalized Canonical Discriminant and Canonical Correlation Analysis. R package version 0.7-0. http://CRAN.R-project.org/package=candisc

Hilbe, J.M. (2009). *Logistic Regression Models*. Boca Raton, FL: Chapman and Hall/ CRC.

Venables, W.N. and Ripley, B.D. (2002). *Modern Applied Statistics*. 4th Edn. New York: Springer.

Weihs, C., Ligges, U., Luebke, K., and Raabe, N. (2005). klaR analyzing German business cycles. In Baier, D., Decker, R., and Schmidt-Thieme, L. (eds). *Data Analysis and Decision Support*, pp. 335–43. Berlin: Springer.

chapter nine

Cluster analysis

9.1   Uses of cluster analysis

Suppose that there is a sample of n objects, each of which has a score on p variables. Then, the idea of a cluster analysis is to use the values of the variables to devise a scheme for grouping the objects into classes so that similar objects are in the same class. The method used must be completely numerical, and the number of classes is not usually known. This problem is clearly more difficult than the problem for a discriminant function analysis that was considered in the last chapter, because with discriminant function analysis, the groups are known to begin with.

There are many reasons why cluster analysis may be worthwhile. It might be a question of finding the true groups that are assumed to really exist. For example, in psychiatry, there has been disagreement over the classification of depressed patients, and cluster analysis has been used to define objective groups. Cluster analysis may also be useful for data reduction. For example, a large number of cities can potentially be used as test markets for a new product, but it is only feasible to use a few. If cities can be placed into a small number of groups of similar cities, then one member from each group can be used for the test market. Alternatively, if cluster analysis generates unexpected groupings, then this might in itself suggest relationships to be investigated.

9.2   Types of cluster analysis

Many algorithms have been proposed for cluster analysis. Here, attention will mostly be restricted to those following two particular approaches. First, there are hierarchic techniques that produce a dendrogram, as shown in Figure 9.1. These methods start with the calculation of the distances from each object to all other objects. Groups are then formed by a process of agglomeration or division. With agglomeration, all objects start by being alone in groups of one. Close groups are then gradually merged, till finally all objects are in a single group. With division, all objects start in a single group. This is then split into two groups, and the two groups are then split, and so on, till all objects are in groups of their own.

The second approach to cluster analysis involves partitioning, with objects being allowed to move in and out of groups at different stages of the analysis. There are many variations on the algorithm used, but

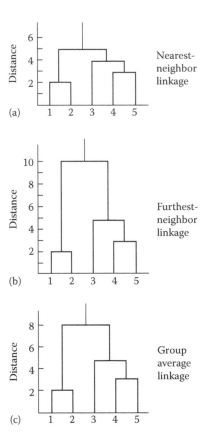

Figure 9.1 (a–c) Examples of dendrograms from cluster analyses of five objects.

the basic approach involves first choosing some more or less arbitrary group centers, with objects then allocated to the nearest one. New centers are then calculated, where these represent the averages of the objects in the groups. An object is then moved to a new group if it is closer to that group's center than it is to the center of its present group. Any groups that are close together are merged, spread-out groups are split, and so on, following some defined rules. The process continues iteratively till stability is achieved with a predetermined number of groups. Usually, a range of values is tried for the final number of groups.

9.3   *Hierarchic methods*

Agglomerative hierarchic methods start with a matrix of distances between objects. All objects begin alone in groups of size one, and groups that are close together are merged. There are various ways to define *close*. The simplest is in terms of nearest neighbors. For example, suppose that

Table 9.1 A matrix showing the distances between five objects

Object	Object				
	1	2	3	4	5
1	—				
2	2	—			
3	6	5	—		
4	10	9	4	—	
5	9	8	5	3	—

Note: The distance is always zero between an object and itself, and the distance from object i to object j is the same as the distance from object j to object i.

Table 9.2 Merging of groups based on nearest-neighbor distances

Distance	Groups
0	1, 2, 3, 4, 5
2	(1,2), 3, 4, 5
3	(1,2), 3, (4,5)
4	(1,2), (3,4,5)
5	(1,2,3,4,5)

the distances between five objects are as shown in Table 9.1. The calculations are then as shown in Table 9.2.

Groups are merged at a given level of distance if one of the objects in one group is at that distance or closer to at least one object in the second group. At a distance of 0, all five objects are on their own. The smallest distance between two objects is 2, which is between the first and second objects. Hence, at a distance level of 2, there are four groups: (1, 2), (3), (4), and (5). The next smallest distance between objects is 3, which is between objects 4 and 5. Hence, at a distance of 3, there are three groups: (1, 2), (3), and (4, 5). The next smallest distance is 4, which is between objects 3 and 4. Hence, at this level of distance, there are two groups: (1,2) and (3,4,5). Finally, the next smallest distance is 5, which is between objects 2 and 3 and between objects 3 and 5. At this level, the two groups merge into the single group (1, 2, 3, 4, 5), and the analysis is complete. The dendrogram shown in Figure 9.1a illustrates how the agglomeration takes place.

With furthest-neighbor linkage, two groups merge only if the most distant members of the two groups are close enough. With the example data, this works as shown in Table 9.3.

Table 9.3 Merging of
groups based on furthest-
neighbor distances

Distance	Groups
0	1, 2, 3, 4, 5
2	(1,2), 3, 4, 5
3	(1,2), 3, (4,5)
5	(1,2), (3,4,5)
10	(1,2,3,4,5)

Table 9.4 Merging of
groups based on group
average distances

Distance	Groups
0	1, 2, 3, 4, 5
2	(1,2), 3, 4, 5
3	(1,2), 3, (4,5)
4.5	(1,2), (3,4,5)
7.8	(1,2,3,4,5)

Object 3 does not join with objects 4 and 5 till distance level 5, because this is the distance to object 3 from the further away of objects 4 and 5. The furthest-neighbor dendrogram is shown in Figure 9.1b.

With group average linkage, two groups merge if the average distance between them is small enough. With the example data, this gives the results shown in Table 9.4. For example, groups (1,2) and (3,4,5) merge at distance level 7.8, as this is the average distance from objects 1 and 2 to objects 3, 4 and 5, the actual distances being 1–3, 6; 1–4, 10, 1–5, 9; 2–3, 5; 2–4, 9; 2–5, 8, with $(6 + 10 + 9 + 5 + 9 + 8)/6 = 7.8$. The dendrogram in this case is shown in Figure 9.1c.

Divisive hierarchic methods have been used less often than agglomerative ones. The objects are all put into one group initially, and then this is split into two groups by separating off the object that is furthest on average from the other objects. Objects from the main group are then moved to the new group if they are closer to this group than they are to the main group. Further subdivisions occur as the distance that is allowed between objects in the same group is reduced. Eventually, all objects are in groups of size one.

9.4   *Problems with cluster analysis*

It has already been mentioned that there are many algorithms for cluster analysis. However, there is no generally accepted best method. Unfortunately, different algorithms do not necessarily produce the same

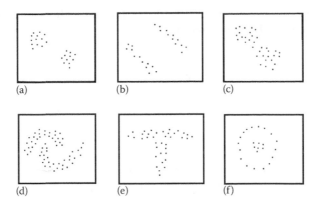

Figure 9.2 (a–f) Some possible patterns of points when there are two clusters.

results on a given set of data, and there is usually rather a large subjective component in the assessment of the results from any particular method.

A fair test of any algorithm is to take a set of data with a known group structure and see whether the algorithm is able to reproduce this structure. It seems to be the case that this test only works in cases where the groups are very distinct. When there is a considerable overlap between the initial groups, a cluster analysis may produce a solution that is quite different from the true situation.

In some cases, difficulties will arise because of the shape of clusters. For example, suppose that there are two variables X_1 and X_2, and objects are plotted according to their values for these. Some possible patterns of points are illustrated in Figure 9.2. Case (a) is likely to be found by any reasonable algorithm, as is case (b). In case (c), some algorithms might well fail to detect two clusters because of the intermediate points. Most algorithms would have trouble handling cases like (d), (e), and (f).

Of course, clusters can only be based on the variables that are given in the data. Therefore, they must be relevant to the classification wanted. To classify depressed patients, there is presumably not much point in measuring height, weight, or length of arms. A problem here is that the clusters obtained may be rather sensitive to the particular choice of variables that is made. A different choice of variables, apparently equally reasonable, may give different clusters.

9.5   Measures of distance

The data for a cluster analysis usually consist of the values of p variables $X_1, X_2, \ldots, X_p$ for n objects. For hierarchic algorithms, these values are then used to produce an array of distances between the objects. Measures of distance have already been discussed in Chapter 5. Here it suffices to say that the Euclidean distance function

$$d_{ij} = \left\{ \sum_{k=1}^{p} \left(x_{ik} - x_{jk} \right)^2 \right\}^{1/2} \tag{9.1}$$

is often used, where:

   x_{ik} is the value of variable X_k for object i
   x_{jk} is the value of the same variable for object j

The geometrical interpretation of the distance d_{ij} is illustrated in Figures 5.1 and 5.2 for the cases of two and three variables.

Usually, variables are standardized in some way before distances are calculated, so that all p variables are equally important in determining these distances. This can be done by coding the variables so that the means are all zero and the variances are all one. Alternatively, each variable can be coded to have a minimum of zero and a maximum of one. Unfortunately, standardization has the effect of minimizing group differences, because if groups are separated well by the variable X_i, then the variance of this variable will be large. In fact, it should be large. It would be best to be able to make the variances equal to one within clusters, but this is obviously not possible, as the whole point of the analysis is to find the clusters.

9.6   *Principal components analysis with cluster analysis*

Some cluster analysis algorithms begin by doing a principal component analysis to reduce a large number of original variables down to a smaller number of principal components. This can drastically reduce the computing time for the cluster analysis. However, it is known that the results of a cluster analysis can be rather different with and without the initial principal components analysis. Consequently, an initial principal components analysis is probably best avoided, because computing time is seldom an issue in the present day.

On the other hand, when the first two principal components account for a high percentage of variation in the data, a plot of individuals against these two components is certainly a useful way for looking for clusters. For example, Figure 6.2 shows European countries plotted in this way for principal components based on employment percentages. The countries do seem to group in a meaningful way.

Example 9.1: Clustering of European countries

The data just mentioned on the percentages of people employed in nine industry groups in different countries of Europe (Table 1.5) can be used for a first example of cluster analysis. The analysis should show which countries have similar employment patterns and which countries are different in this respect. As shown in Table 1.5, a grouping existed when the data were collected, consisting of (1) the European Union (EU) countries, (2) the European Free Trade Area (EFTA) countries, (3) the Eastern European countries, and (4) the four other countries of Cyprus, Gibraltar, Malta, and Turkey. It is, therefore, interesting to see whether this grouping can be recovered using a cluster analysis.

The first step in the analysis involves standardizing the nine variables so that each one has a mean of zero and a standard deviation of one. For example, variable 1 is AGR, the percentage employed in agriculture, forestry, and fishing. For the 30 countries being considered, this variable has a mean of 12.19 and a standard deviation of 12.31, with the latter value calculated using Equation 4.1. The AGR data value for Belgium is 2.6, which standardizes to $(2.6 - 12.19)/12.31 = -0.78$. Similarly, the data value for Denmark is 5.6, which standardizes to -0.54, and so on. The standardized data values are shown in Table 9.5.

The next step in the analysis involves calculating the Euclidean distances between all pairs of countries. This can be done by applying Equation 9.1 to the standardized data values. Finally, a dendrogram can be formed using, for example, the agglomerative, nearest-neighbor, hierarchic process described in Section 9.3. In practice, all these steps can be carried out using a suitable statistical package.

The dendrogram obtained using the NCSS package (Hintze, 2012) is shown in Figure 9.3. It can be seen that the two closest countries are Sweden and Denmark. These are at a distance of about 0.2 apart. At a slightly larger distance, Belgium joins these two countries to make a cluster of size three. As the distance increases, more and more countries combine, and the amalgamation ends with Albania joining all the other countries in one cluster, at a distance of about 1.7.

One interpretation of the dendrogram is that there are just four clusters defined by a nearest-neighbor distance of about 1.0. These are (1) Albania, (2) Hungary and the Czech/Slovak Republic, (3) Gibraltar, and (4) all the other countries. This, then, separates off three eastern countries and Gibraltar from everything else and suggests that the classification into EU, EFTA, eastern, and other countries is not a good indicator of employment patterns. This contradicts the reasonably successful separation of eastern and EU/EFTA countries by a discriminant function analysis (Figure 8.1). However, there is some limited agreement with the plot of countries against the first two principal component, where Albania and Gibraltar show up as having very extreme data values (Figure 6.2).

An alternative analysis was carried out using the K-means clustering option in the NCSS package (Hintze, 2012). This essentially

Table 9.5 Standardized values for percentages employed in different industry groups in Europe, derived from the percentages in Table 5.1

Country	AGR	MIN	MAN	PS	CON	SER	FIN	SPS	TC
Belgium	−0.78	−0.37	0.05	0.00	−0.45	0.24	0.51	1.13	0.28
Denmark	−0.54	−0.38	0.01	−0.16	−0.41	−0.22	0.61	1.07	0.44
France	−0.58	−0.35	−0.01	0.16	−0.16	0.21	0.89	0.70	−0.04
Germany	−0.73	−0.31	0.48	0.32	0.68	0.30	0.74	0.16	−0.69
Greece	0.81	−0.33	−0.11	0.32	−0.27	0.50	−0.34	−0.82	0.36
Ireland	0.13	−0.32	−0.05	0.64	−0.16	0.42	0.44	−0.17	−0.53
Italy	−0.31	−0.26	0.17	−1.29	0.57	1.16	−0.51	0.12	−0.94
Luxembourg	−0.72	−0.38	−0.07	−0.16	0.87	1.08	0.51	0.30	0.28
The Netherlands	−0.65	−0.38	−0.11	−0.16	−2.54	0.55	1.22	1.29	0.28
Portugal	−0.06	−0.33	0.35	−0.16	0.25	0.81	−0.09	−0.27	−1.34
Spain	−0.19	−0.33	0.09	−0.32	0.72	0.86	−0.19	−0.03	−0.53
United Kingdom	−0.81	−0.31	0.11	0.64	−0.19	0.88	1.44	0.16	0.04
Austria	−0.39	−0.35	0.70	0.64	0.35	0.67	0.01	−0.42	−0.04
Finland	−0.30	−0.37	−0.10	0.64	−0.27	−0.20	0.49	0.71	0.85
Iceland	−0.14	−0.39	−0.17	0.16	0.90	−0.22	0.34	0.42	0.20
Norway	−0.52	−0.26	−0.60	0.48	−0.38	0.38	0.24	1.20	1.34
Sweden	−0.73	−0.35	−0.14	0.00	−0.41	−0.28	0.69	1.43	0.61
Switzerland	−0.54	−0.39	0.47	−1.29	0.61	0.94	1.02	−0.45	−0.21
Albania	3.52	1.80	−2.15	−1.29	−1.51	−2.39	2.17	−3.09	−2.80
Bulgaria	0.55	−0.39	1.56	−1.29	−0.30	−1.21	−1.29	−0.70	0.85
Czech/Slovak Reps	0.05	3.82	−2.15	−1.29	0.32	−1.05	−1.27	−0.47	0.36
Hungary	0.25	2.87	−2.15	−1.29	−0.41	−0.45	−1.67	0.04	1.90
Poland	0.93	0.05	0.40	0.16	−0.45	−1.03	−1.34	−0.29	−1.02
Romania	0.80	−0.10	1.86	1.93	−0.63	−1.69	−1.52	−1.34	0.28
USSR (former)	0.51	−0.39	0.90	−1.29	0.98	−1.50	−1.52	−0.16	1.58
Yugoslavia (former)	−0.58	−0.14	1.95	2.25	0.21	−0.36	−0.89	−0.90	1.09
Cyprus	0.11	−0.35	−0.14	−0.48	0.57	1.56	0.01	−0.66	−0.37
Gibraltar	−0.99	−0.39	−1.43	1.93	3.43	1.72	1.04	0.80	−1.18
Malta	−0.78	−0.32	0.81	1.13	−1.07	−1.05	−0.69	1.67	0.61
Turkey	2.65	−0.29	−0.53	−0.97	−0.85	−0.63	−1.07	−1.43	−1.66

uses the partitioning method described in Section 9.2, which starts with arbitrary cluster centers, allocates items to the nearest center, recalculates the mean values of variables for each group, reallocates individuals to their closest group centers to minimize the within-cluster total sum of squares, and so on. The calculations use standardized variables with means of zero and standard deviations of one. Ten random choices of starting clusters were tried, with from two to six clusters.

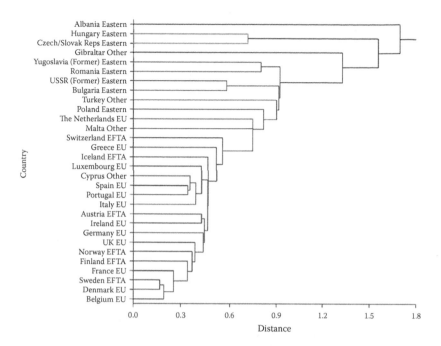

Figure 9.3 The dendrogram obtained from a nearest-neighbor, hierarchic cluster analysis on employment data from European countries.

The percentage of the variation accounted for varied from 73.4% with two clusters to 27.6% with six clusters. With four clusters, these were (1) Turkey and Albania, (2) Hungary and the Czech/Slovak Republics, (3) Bulgaria, Poland, Romania, the USSR (former), Yugoslavia (former), and Malta, and (4) the EU and EFTA countries, Cyprus, and Gibraltar. This is not the same as the four-cluster solution given by the dendrogram of Figure 9.3, although there are some similarities. No doubt, other algorithms for cluster analysis will give slightly different solutions.

Example 9.2: Relationships between canine species

As a second example, consider the data provided in Table 1.4 for mean mandible measurements of seven canine groups. As has been explained in Example 1.4 in Chapter 1, these data were originally collected as part of a study on the relationship between prehistoric dogs, whose remains have been uncovered in Thailand, and the other six living species. This question has already been considered in terms of distances between the seven groups in Example 5.1. Table 5.1 shows mandible measurements standardized to have means of zero and standard deviations of one. Table 5.2 shows Euclidean distances between the groups based on these standardized measurements.

Table 9.6 Clusters found at different distance levels for a hierarchic
nearest-neighbor cluster analysis

Distance	Clusters	Number of clusters
0.00	MD, PD, GJ, CW, IW, CU, DI	7
0.72	(MD, PD), GJ, CW, IW, CU, DI	6
1.38	(MD, PD, CU), GJ, CW, IW, DI	5
1.63	(MD, PD, CU), GJ, CW, IW, DI	5
1.68	(MD, PD, CU, DI), GJ, CW, IW	4
1.80	(MD, PD, CU, DI), GJ, CW, IW	4
1.84	(MD, PD, CU, DI), GJ, CW, IW	4
2.07	(MD, PD, CU, DI, GJ), CW, IW	3
2.31	(MD, PD, CU, DI, GJ), (CW, IW)	2

Note: MD = modern dog, GJ = golden jackal, CW = Chinese wolf,
IW = Indian wolf, CU = cuon, DI = dingo and PD = prehistoric dog.

With only seven species to cluster, it is a simple matter to carry out a nearest-neighbor, hierarchic cluster analysis without using a computer. Thus, it can be seen from Table 5.2 that the two most similar species are the prehistoric dog and the modern dog, at a distance of 0.72. These, therefore join, into a single cluster at that level. The next largest distance is 1.38 between the cuon and the prehistoric dog, so that at that level, the cuon joins the cluster with the prehistoric and modern dogs. The third largest distance is 1.63 between the cuon and the modern dog, but because these are already in the same cluster, this has no effect. Continuing in this way produces the clusters that are shown in Table 9.6. The corresponding dendrogram is shown in Figure 9.4.

It appears that the prehistoric dog is closely related to the modern Thai dog, with both of these being somewhat related to the cuon and dingo and less closely related to the golden jackal. The Indian and Chinese wolves are closest to each other, but the difference between them is relatively large.

It seems fair to say that in this example, the cluster analysis has produced a sensible description of the relationship between the different groups.

9.7   *Computer programs*

Computer programs for cluster analysis are widely available, and the larger statistical packages often include a variety of different options for both hierarchic and partitioning methods. As the results obtained usually vary to some extent depending on the precise details of the algorithms used, it will usually be worthwhile to try several options before deciding on the final method to be used for an analysis. Doing the calculations using methods available in R packages is described in the Appendix to this chapter.

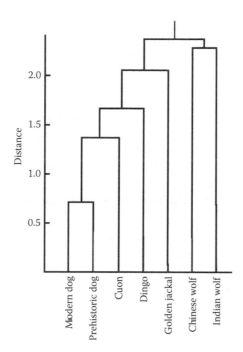

Figure 9.4 Dendrogram produced from a nearest-neighbor cluster analysis for the relationship between canine species.

9.8   *Discussion and further reading*

A number of books devoted to cluster analysis are available, including the classic texts by Hartigan (1975), Romesburg (2004), and Everitt et al. (2011).

An approach to clustering that has not been considered in this chapter involves assuming that the data available come from a mixture of several different populations for which the distributions are of a type that is assumed to be known (e.g., multivariate normal). The clustering problem is then transformed into the problem of estimating, for each of the populations, the parameters of the assumed distribution and the probability that an observation comes from that population. This approach has the merit of moving the clustering problem away from the development of ad hoc procedures toward the more usual statistical framework of parameter estimation and model testing. See Everitt et al. (2011, chapter 6) for an introduction to this method.

Exercises

1. Table 9.7 shows the abundances of the 25 most abundant plant species on 17 plots from a grazed meadow in Steneryd Nature Reserve in Sweden, as measured by Persson (1981) and used for an example

by Digby and Kempton (1987). Each value in the table is the sum of cover values in a range from 0 to 5 for nine sample quadrats, so that a value of 45 corresponds to complete cover by the species being considered. Note that the species are in order from the most abundant (1) to the least abundant (25), and the plots are in the order given by Digby and Kempton (1987, table 3.2), which corresponds to variation in certain environmental factors such as light and moisture. Carry out cluster analyses to study the relationships between (a) the 17 plots, and (b) the 25 species.

2. Table 9.8 shows a set of data concerning grave goods from a cemetery at Bannadi, northeast Thailand, which were kindly supplied by Professor C.F.W. Higham. These data consist of a record of the

Table 9.7 Abundance measures for 25 plant species on 17 plots in Steneryd Nature Reserve, Sweden

Species	\	\	\	\	\	\	\	Plot	\	\	\	\	\	\	\	\	\
	1	2	3	4	5	6	7	8	9	10	11	12	13	14	15	16	17
Festuca ovina	38	43	43	30	10	11	20	0	0	5	4	1	1	0	0	0	0
Anemone nemorosa	0	0	0	4	10	7	21	14	13	19	20	19	6	10	12	14	21
Stallaria holostea	0	0	0	0	0	6	8	21	39	31	7	12	0	16	11	6	9
Agrostis tenuis	10	12	19	15	16	9	0	9	28	8	0	4	0	0	0	0	0
Ranunculus ficaria	0	0	0	0	0	0	0	0	0	0	13	0	0	21	20	21	37
Mercurialis perennis	0	0	0	0	0	0	0	0	0	0	1	0	0	0	11	45	45
Poa pratensis	1	0	5	6	2	8	10	15	12	15	4	5	6	7	0	0	0
Rumex acetosa	0	7	0	10	9	9	3	9	8	9	2	5	5	1	7	0	0
Veronica chamaedrys	0	0	1	4	6	9	9	9	11	11	6	5	4	1	7	0	0
Dactylis glomerata	0	0	0	0	0	8	0	14	2	14	3	9	8	7	7	2	1
Fraxinus excelsior (juv.)	0	0	0	0	0	8	0	0	6	5	4	7	9	8	8	7	6
Saxifraga granulate	0	5	3	9	12	9	0	1	7	4	5	1	1	1	3	0	0
Deschampsia flexuosa	0	0	0	0	0	0	30	0	14	3	8	0	3	3	0	0	0
Luzula campestris	4	10	10	9	7	6	9	0	0	2	1	0	2	0	1	0	0
Plantago lanceolata	2	9	7	15	13	8	0	0	0	0	0	0	0	0	0	0	0
Festuca rubra	0	0	0	0	15	6	0	18	1	9	0	0	2	0	0	0	0
Hieracium pilosella	12	7	16	8	1	6	0	0	0	0	0	0	0	0	0	0	0
Geum urbanum	0	0	0	0	0	7	0	2	2	1	0	7	9	2	3	8	7
Lathyrus montanus	0	0	0	0	0	7	9	2	12	6	3	8	0	0	0	0	0
Campanula persicifolia	0	0	0	0	2	6	3	0	6	5	3	9	3	2	7	0	0
Viola riviniana	0	0	0	0	0	4	1	4	2	9	6	8	4	1	6	0	0
Hepatica nobilis	0	0	0	0	0	8	0	4	0	6	2	10	6	0	2	7	0
Achillea millefolium	1	9	16	9	5	2	0	0	0	0	0	0	0	0	0	0	0
Allium sp.	0	0	0	0	2	7	0	1	0	3	1	6	8	2	0	7	4
Trifolim repens	0	0	6	14	19	2	0	0	0	0	0	0	0	0	0	0	0

Table 9.8 Grave goods in burials in the Bannadi Cemetery in northern Thailand

Burial	Type	1	2	3	4	5	6	7	8	9	10	11	12	13	14	15	16	17	18	19	20	21	22	23	24	25	26	27	28	29	30	31	32	33	34	35	36	37	38	Sum
B33	3	0	0	0	0	0	0	0	0	0	0	0	0	0	0	0	0	0	0	0	0	0	0	0	0	0	0	0	0	0	0	0	0	0	0	0	0	0	0	0
B9	2	0	0	0	0	0	0	0	0	0	0	0	0	0	0	0	0	0	0	0	0	0	0	0	0	0	0	0	0	0	0	0	0	0	0	0	0	0	0	0
B32	2	0	0	0	0	0	0	0	0	0	0	0	0	0	0	0	0	0	0	0	0	0	0	0	0	0	0	0	0	0	0	0	0	0	0	0	0	0	0	0
B11	1	0	0	0	0	0	0	0	0	0	0	0	0	0	0	0	0	0	0	0	0	0	0	0	0	0	0	0	0	0	0	0	0	0	0	0	0	0	0	0
B28	1	0	0	0	0	0	0	0	0	0	0	0	0	0	0	0	0	0	0	0	0	0	0	0	0	0	0	0	0	0	0	0	0	0	0	0	0	0	0	0
B41	2	0	0	0	0	0	0	0	0	0	0	0	0	0	0	0	0	0	0	0	0	0	0	0	0	0	0	0	0	0	0	0	0	0	0	0	0	0	0	0
B27	2	0	0	0	0	0	0	0	0	0	0	0	0	0	0	0	0	0	0	0	0	0	0	0	0	0	0	0	0	0	0	0	0	0	0	0	0	0	0	0
B24	2	0	0	0	0	0	0	0	0	0	0	0	0	0	0	0	0	0	0	0	0	0	0	0	0	0	0	0	0	0	0	0	0	0	0	0	0	0	0	0
B39	1	0	0	0	0	0	0	0	0	0	0	0	0	0	0	0	0	0	0	0	0	0	0	0	0	0	0	0	0	0	0	0	0	0	0	0	0	0	0	0
B43	2	0	0	0	0	0	0	0	0	0	0	0	0	0	0	0	0	0	0	0	0	0	0	0	0	0	0	0	0	0	0	0	0	0	0	0	0	0	0	0
B20	2	0	0	0	0	0	0	0	0	0	0	0	0	0	0	0	0	0	0	0	0	0	0	0	0	0	0	0	0	0	0	0	0	0	0	0	0	0	0	0
B34	3	0	0	0	0	0	0	0	0	0	0	0	0	0	0	0	0	0	0	0	0	0	0	0	0	0	0	0	0	0	0	0	0	0	0	0	0	0	1	1
B27	1	0	0	0	0	0	0	0	0	0	0	0	0	0	0	0	0	0	0	0	0	0	0	0	0	0	0	0	0	0	0	0	0	0	0	0	0	1	0	1
B37	1	0	0	0	0	0	0	1	0	0	0	0	0	0	0	0	0	0	0	0	0	0	0	0	0	0	0	0	0	0	0	0	0	0	0	0	0	0	0	1
B25	2	0	0	0	0	0	0	0	0	0	0	0	0	0	0	0	0	0	0	0	0	0	0	0	0	0	0	0	0	0	0	1	0	0	0	0	0	0	0	1
B30	2	0	0	0	0	0	0	0	0	0	0	0	0	0	0	0	0	0	0	0	0	0	0	0	0	0	0	0	0	0	0	0	0	0	1	0	0	0	0	1
B21	1	0	0	0	0	0	0	0	0	0	0	0	0	0	0	0	0	0	0	0	0	0	0	0	0	0	0	0	0	0	0	0	0	1	0	0	0	0	0	1
B49	2	0	0	0	0	0	0	0	0	0	0	0	0	0	0	0	0	0	0	0	0	0	0	0	0	0	0	0	0	1	0	0	0	0	0	0	0	0	0	1
B40	2	0	0	0	0	0	0	0	0	0	0	0	0	0	0	0	0	0	0	0	0	0	0	0	0	0	0	0	1	1	0	0	0	0	0	0	0	0	0	2
BT8	2	0	0	0	0	0	0	0	0	0	0	0	0	0	0	0	0	0	0	0	0	0	0	0	0	0	0	0	0	0	0	0	0	0	0	1	1	0	0	2
BT17	2	0	0	0	0	0	0	0	0	0	0	0	0	0	0	0	0	0	0	0	0	0	0	0	0	0	0	0	0	0	0	0	0	0	0	0	1	1	0	2
BT21	1	0	0	0	0	0	0	0	0	0	0	0	0	0	0	0	0	0	2	0	0	0	0	0	0	0	0	0	0	0	0	0	0	0	0	0	0	0	0	2
BT5	1	0	0	0	0	0	0	0	0	0	0	0	0	0	0	0	0	0	3	0	0	0	0	0	0	0	0	0	0	0	0	0	0	0	0	0	0	0	0	3
B14	3	0	0	0	0	0	0	0	0	0	0	0	0	0	0	0	0	0	0	0	0	0	0	0	0	0	0	0	0	0	0	0	0	0	1	1	0	1	0	3
B31	1	0	0	0	0	0	0	0	0	0	0	0	0	0	0	0	0	0	0	0	0	0	0	0	0	0	0	0	0	0	0	0	1	0	1	0	0	1	0	3
B42	1	0	0	0	0	0	0	0	0	0	0	0	0	0	0	0	0	0	0	0	0	0	1	0	0	0	0	0	0	0	0	0	0	0	0	0	0	1	1	3

(Continued)

Table 9.8 (Continued) Grave goods in burials in the Bannadi Cemetery in northern Thailand

Burial	Type	1	2	3	4	5	6	7	8	9	10	11	12	13	14	15	16	17	18	19	20	21	22	23	24	25	26	27	28	29	30	31	32	33	34	35	36	37	38	Sum
B44	2	0	0	0	0	0	0	0	0	0	0	0	0	0	0	0	0	0	0	0	0	0	0	0	0	0	0	0	0	0	0	1	0	1	0	0	1	0	0	3
B35	1	0	0	0	0	0	0	0	1	0	0	0	0	0	0	0	0	0	0	0	0	0	0	0	0	0	0	0	0	1	0	0	0	1	0	0	0	0	0	3
BT15	1	0	0	0	0	0	0	0	0	0	0	0	0	0	0	0	0	0	0	0	0	0	0	0	0	0	0	0	1	0	0	0	0	0	0	0	1	1	1	3
B15	3	1	0	0	0	0	0	0	0	0	0	0	0	0	0	0	0	0	0	0	0	0	0	0	0	0	0	0	0	0	0	0	0	0	0	1	1	1	1	4
B45	3	0	0	0	0	0	0	0	0	0	0	0	0	0	0	0	0	0	0	0	0	0	0	0	0	0	0	0	1	0	0	0	0	1	0	1	0	1	0	4
B46	3	0	0	0	0	0	0	0	0	0	0	0	0	0	0	0	0	0	0	0	0	0	0	0	0	0	0	0	0	1	0	0	0	1	0	0	1	0	0	4
B17	1	0	0	0	0	0	0	0	0	0	0	0	0	1	0	0	0	0	0	0	0	0	0	0	0	0	0	0	0	0	1	0	0	1	0	0	0	0	1	4
B10	2	0	0	0	0	0	0	0	0	0	0	0	0	0	0	0	0	1	0	0	0	0	1	0	0	0	0	1	0	0	0	0	0	1	0	1	0	1	0	4
BT16	2	0	0	0	0	0	0	0	0	0	0	0	0	0	0	0	0	0	0	0	0	0	0	0	0	0	0	0	0	0	1	0	0	1	0	1	0	1	1	4
B26	2	0	0	0	0	0	0	0	0	0	0	0	0	0	0	0	0	0	0	0	0	0	0	0	0	0	0	1	1	0	0	0	0	0	0	1	1	1	1	4
B16	1	0	1	0	1	0	0	0	0	0	0	0	0	0	0	0	0	0	0	0	0	1	0	0	0	0	0	0	1	0	0	0	0	0	0	1	0	0	0	5
B29	3	0	0	0	0	0	0	0	0	0	0	0	0	0	0	1	1	0	0	0	0	0	0	0	0	0	0	0	0	0	0	0	0	0	0	1	1	0	1	5
B19	3	0	0	0	0	0	0	0	0	0	1	1	0	0	0	0	0	0	0	0	0	0	0	0	1	1	0	0	0	0	0	0	0	1	0	0	0	0	0	6
B32	2	0	0	0	0	0	0	0	0	0	0	0	0	0	0	0	0	0	0	0	0	0	0	0	1	0	1	0	0	0	1	1	1	0	1	0	0	0	1	6
B38	3	0	0	0	0	0	0	0	0	0	0	0	0	0	0	0	0	0	0	0	0	0	0	0	1	0	0	0	1	1	1	1	0	1	0	0	1	0	1	7
B36	2	0	0	0	0	0	0	0	0	0	0	0	0	0	1	0	0	0	0	0	0	0	0	0	0	0	0	1	0	0	1	0	0	1	1	1	1	1	1	7
B12	2	0	0	0	0	0	0	0	0	0	1	1	0	0	0	0	0	0	0	0	0	0	0	1	0	1	1	1	0	0	0	0	1	1	1	0	0	0	0	8
BT12	1	0	0	0	0	0	0	0	0	0	0	0	0	0	0	0	0	0	0	0	0	1	0	0	0	0	0	0	0	1	1	0	1	1	1	1	1	1	1	8
B47	1	0	0	1	0	1	0	0	0	0	0	0	1	0	0	0	0	0	1	1	0	0	1	0	0	0	0	0	1	1	0	0	0	0	0	0	0	1	8
B18	2	0	0	0	0	1	1	0	0	0	0	0	0	0	0	0	0	0	1	1	0	0	0	1	0	1	0	0	0	0	1	0	0	0	0	0	0	1	1	9
B48	2	0	0	0	1	1	1	1	0	1	0	0	0	0	0	0	1	1	0	1	0	0	0	0	0	0	1	0	0	0	0	0	0	0	0	1	1	1	1	11
Sum		1	1	1	1	1	1	1	1	1	1	1	1	1	1	1	1	1	1	1	1	1	2	2	3	3	3	3	4	6	6	6	7	8	9	12	15	16	18	144

Note: Body types: 1, adult male; 2, adult female; 3, child.

presence or absence of 38 different types of article in each of 47 graves, with additional information on whether the body was of an adult male, an adult female, or a child. The burials are in the order of richness of the different types of goods (from 0 to 11), and the goods are in the order of the frequency of occurrence (from 1 to 18). Carry out a cluster analysis to study the relationships between the 47 burials. Is there any clustering in terms of the type of body?

References

Digby, P.G.N. and Kempton, R.A. (1987). *Multivariate Analysis of Ecological Communities.* London: Chapman and Hall.

Everitt, B., Landau, S., and Leese, M. (2011). *Cluster Analysis.* 5th Edn. New York: Wiley.

Hartigan, J. (1975). *Clustering Algorithms.* New York: Wiley.

Hintze, J. (2012). *NCSS8.* NCSS LLC. www.ncss.com.

Persson, S. (1981). Ecological indicator values as an aid in the interpretation of ordination diagrams. *Journal of Ecology* 69: 71–84.

Romesburg, H.C. (2004). *Cluster Analysis for Researchers.* Morrisville: Lulu.com.

Appendix: Cluster Analysis in R

The information needed to produce agglomerative dendrograms for a set of objects is performed by `hclust()`, an R function that implements the most common algorithms for hierarchical clustering. The first argument of `hclust()` is a distance or dissimilarity matrix (of objects), like any of those generated by the function `dist()`, as described in the Appendix for Chapter 5. The default algorithm in `hclust` is the furthest-neighbor or complete linkage (`method=complete`). Other agglomerative methods available include those described in Chapter 9 (single or nearest-neighbor and average linkage) and extra methods indicated in the corresponding help documentation of `hclust`. As an example, given a distance matrix `d.mat`, the clustering of objects by means of the group average linkage will be written as

```
gr.av.link <- hclust(d.mat, average)
```

The resulting object contains crucial information about the clustering process. That object belongs to the class `hclust`, and it can be used as the main argument for plotting a dendrogram. Using

```
plot(gr.av.link)
```

will produce a dendrogram with labels for each object hanging exactly where a branch starts, not necessarily at 0 distance. The branches of dendrograms displayed in Chapter 9 all start at 0, a situation that is forced by assigning a negative number after the option `hang=`. For example,

```
plot(gr.av.link, hang = -1)
```

generates a dendrogram of `hclust` object `gr.av.link` with all the labels hanging down from 0. To gain further control of a dendrogram's appearance using the plot function, it is necessary to convert an `hclust` object into a dendrogram object with the command `as.dendrogram()`. As an example, the following command will display a horizontal dendrogram based on the contents of the `hclust` object generated above, with all object labels starting next to the 0 distance:

```
plot(as.dendrogram(gr.av.link), hang = -1, horiz = TRUE)
```

More options are available for `hclust`. One is the command `cutree`, which allows the user to cut a dendrogram (tree) into several groups by specifying either the desired number of groups or the cut heights, that is, the distance at which groups are identifiable. Another useful command is `rect.hclust`, which draws rectangles around the branches of a dendrogram highlighting the corresponding clusters. With this command, first the dendrogram is cut at a certain level, and then a rectangle is drawn around selected branches.

Several computational tools have been implemented in the standard R and special cluster analysis packages for data mining, that is, looking for patterns in the data. Thus, clustering techniques have been put together in the more extensive package `cluster` (Maechler et al., 2016), which includes functions for hierarchical and nonhierarchical clustering and other methods, all of them with feminine names such as `agnes`, `daisy`, `pam`, and `clara`. As an example, the power of `hclust` has been improved with `agnes()`, an R function located in the `cluster` package that yields extra hierarchical clustering algorithms as well as an agglomerative coefficient measuring the amount of clustering structure found. It also provides a *banner*, which is a graphical display equivalent to a dendrogram. See the details in the help documentation of the `cluster` package. R scripts found at the book's website have been written to produce the dendrograms shown in Figures 9.1 and 9.3 (Example 9.1) and 9.4 (Example 9.2) using `hclust` and `agnes` functions. Another R script has also been included to show the application of the function `kmeans()`, the nonhierarchical clustering method described in Example 9.1.

Reference

Maechler, M., Rousseeuw, P., Struyf, A., Hubert, M., and Hornik, K. (2016). cluster: Cluster Analysis Basics and Extensions. R package version 2.0.4.

chapter ten

Canonical correlation analysis

10.1   Generalizing a multiple regression analysis

In some sets of multivariate data the variables divide naturally into two groups. A canonical correlation analysis can then be used to investigate the relationships between the two groups. A case in point is the data provided in Table 1.3 in which 16 colonies of the butterfly *Euphydryas editha* in California and Oregon are considered. For each colony, values are available for four environmental variables and six gene frequencies. An obvious question to be considered is what relationships, if any, exist between the gene frequencies and the environmental variables. One method to investigate this subject is through a canonical correlation analysis.

Another example was provided by Hotelling (1936) where he described a canonical correlation analysis for the first time. This example involved the results of tests for reading speed (X_1), reading power (X_2), arithmetic speed (Y_1), and arithmetic power (Y_2) for 140 seventh-grade schoolchildren. The specific question that was addressed was whether or not reading ability (as measured by X_1 and X_2) is related to arithmetic ability (as measured by Y_1 and Y_2).

The approach that a canonical correlation analysis takes to answering this question is to search for a linear combination of X_1 and X_2

$$U = a_1 X_1 + a_2 X_2$$

and a linear combination of Y_1 and Y_2

$$V = b_1 Y_1 + b_2 Y_2$$

where these are chosen to make the correlation between U and V as large as possible. This is somewhat similar to the idea behind a principal components analysis, except that here a correlation is maximized instead of a variance.

With X_1, X_2, Y_1, and Y_2 standardized to have unit variances, Hotelling found that the best choices for U and V with the reading and arithmetic example were

$$U = -2.78 X_1 + 2.27 X_2$$

and

$$V = -2.44Y_1 + 1.00Y_2$$

where these two variables have a correlation of 0.62. It can be seen that U measures the difference between reading power and speed, and V measures the difference between arithmetic power and speed. Hence, it appears that that children with a large difference between X_1 and X_2 also tended to have a large difference between Y_1 and Y_2. It is this aspect of reading and arithmetic that shows the most correlation.

In a multiple regression analysis, a single variable Y is related to two or more variables X_1, X_2, ..., X_p, to see how Y is related to the X variables. From this point of view, canonical correlation analysis is a generalization of multiple regression in which several Y variables are simultaneously related to several X variables.

In practice, more than one pair of canonical variables can be calculated from a set of data. If there are p variables X_1, X_2, ..., X_p and q variables Y_1, Y_2, ..., Y_q, then there can be up to the minimum of p and q pairs of variables. That is to say, linear relationships

$$U_1 = a_{11}X_1 + a_{12}X_2 + \cdots + a_{1p}X_p$$
$$U_2 = a_{21}X_1 + a_{22}X_2 + \cdots + a_{2p}X_p$$

$$\cdot$$

$$\cdot$$

$$\cdot$$

$$U_r = a_{r1}X_1 + a_{r2}X_2 + \cdots + a_{rp}X_p$$

and

$$V_1 = b_{11}Y_1 + b_{12}Y_2 + \cdots + b_{1q}Y_q$$
$$V_2 = b_{21}Y_1 + b_{22}Y_2 + \cdots + b_{2q}Y_q$$

$$\cdot$$

$$\cdot$$

$$\cdot$$

$$V_r = b_{r1}Y_1 + b_{r2}Y_2 + \cdots + b_{rq}Y_q$$

can be established, where r is the smaller of p and q. These relationships are chosen so that the correlation between U_1 and V_1 is a maximum; the correlation between U_2 and V_2 is a maximum, subject to these variables being uncorrelated with U_1 and V_1; the correlation between U_3 and V_3 is a maximum, subject to these variables being uncorrelated with U_1, V_1, U_2, and V_2; and so on. Each of the pairs of canonical variables (U_1, V_1), (U_2, V_2), ... , (U_r, V_r) then represents an independent dimension in the relationship between the two sets of variables $(X_1, X_2, ..., X_p)$ and $(Y_1, Y_2, ..., Y_q)$. The first pair (U_1, V_1) have the highest possible correlation and are therefore the most important, the second pair (U_2, V_2) have the second highest correlation and are therefore the second most important, and so on.

10.2   Procedure for a canonical correlation analysis

Assume that the $(p + q) \times (p + q)$ correlation matrix between the variables $X_1, X_2, ..., X_p, Y_1, Y_2, ..., Y_q$ takes the following form when it is calculated from the sample for which the variables are recorded:

$$
\begin{array}{c}
\begin{array}{cccccccc} X_1 & X_2 & ... & X_p & \quad Y_1 & Y_2 & ... & Y_q \end{array} \\
\begin{array}{c} X_1 \\ X_2 \\ \cdot \\ \cdot \\ \cdot \\ X_p \\ Y_1 \\ Y_2 \\ \cdot \\ \cdot \\ \cdot \\ Y_q \end{array}
\left[
\begin{array}{c|c}
\begin{array}{c} p \times p \text{ matrix} \\[1em] A \end{array} & \begin{array}{c} p \times q \text{ matrix} \\[1em] C \end{array} \\
\hline
\begin{array}{c} q \times p \text{ matrix} \\[1em] C' \end{array} & \begin{array}{c} q \times q \text{ matrix} \\[1em] B \end{array}
\end{array}
\right]
\end{array}
$$

From this matrix, a $q \times q$ matrix $\mathbf{B}^{-1}\mathbf{C}'\mathbf{A}^{-1}\mathbf{C}$ can be calculated, and the eigenvalue problem

$$\left(\mathbf{B}^{-1}\mathbf{C}'\mathbf{A}^{-1}\mathbf{C} - \lambda\mathbf{I}\right)\mathbf{b} = \mathbf{0} \tag{10.1}$$

can be considered. It turns out that the eigenvalues $\lambda_1 > \lambda_2 > ... > \lambda_r$ are then the squares of the correlations between the canonical variables, and the corresponding eigenvectors, $\mathbf{b}_1, \mathbf{b}_2, ..., \mathbf{b}_r$, give the coefficients of the Y variables for the canonical variables. Also, the coefficients of U_i, the ith canonical variable for the X variables, are given by the elements of the vector

$$\mathbf{a}_i = \mathbf{A}^{-1}\mathbf{C}\mathbf{b}_i \qquad (10.2)$$

In these calculations it is assumed that the original X and Y variables are in a standardized form with means of zero and standard deviations of unity. The coefficients of the canonical variables are for these standardized variables.

From Equations 10.1 and 10.2, the ith pair of canonical variables are calculated as

$$U_i = \mathbf{a}_i'\mathbf{X}$$

and

$$V_i = \mathbf{b}_i'\mathbf{Y}$$

where:
$$\begin{aligned}
\mathbf{a}_i' &= (a_{i1}, a_{i2}, \ldots, a_{ip}) \\
\mathbf{b}_i' &= (b_{i1}, b_{i2}, \ldots, b_{iq}) \\
\mathbf{X}' &= (x_1, x_2, \ldots, x_p) \\
\mathbf{Y}' &= (y_1, y_2, \ldots, y_q), \text{ with the X and Y values standardized}
\end{aligned}$$

As they stand, U_i and V_i will have variances that depend on the scaling adopted for the eigenvector $\mathbf{b}_i$. However, it is a simple matter to calculate the standard deviation of U_i for the data and divide the a_{ij} values by this standard deviation. This produces a scaled canonical variable U_i with unit variance. Similarly, if the b_{ij} values are divided by the standard deviation of V_i, then this produces a scaled V with unit variance.

This form of standardization of the canonical variables is not essential, because the correlation between U_i and V_i is not affected by scaling. However, it may be useful when it comes to examining the numerical values of canonical variables for the individuals for which data are available.

10.3   Tests of significance

An approximate test for a relationship between the X variables as a whole and the Y variables as a whole was proposed by Bartlett (1947) for the situation where the data are a random sample from a multivariate normal distribution. This involves calculating the statistic

$$X^2 = -\left\{n - \tfrac{1}{2}(p+q+3)\right\} \sum_{i=1}^{r} \log_e(1-\lambda_i) \qquad (10.3)$$

where n is the number of cases for which data are available. The statistic can be compared with the percentage points of the chi-squared distribution with pq degrees of freedom (df), and a significantly large value provides evidence that at least one of the r canonical correlations is significant. A nonsignificant result indicates that even the largest canonical correlation can be accounted for by sampling variation only.

It is sometimes suggested that this test can be extended to allow the importance of each of the canonical correlations to be tested. Common suggestions are to

1. Compare the ith contributions, $-\{ n-\tfrac{1}{2}(p + q + 3)\}\log_e(1-\lambda_i)$, to the right-hand side of Equation 10.3 with the percentage points of the chi-squared distribution with $p + q - 2i + 1$ df
2. Compare the sum of the (i + 1)th to the rth contributions toward the sum on the right-hand side of Equation 10.3) with the percentage points of the chi-squared distribution with $(p - i)(q - i)$ df.

Here, the first approach is assumed to be testing the ith canonical correlation directly, whereas the second is assumed to be testing for the significance of the (i + 1)th to rth canonical correlations as a whole.

The reason why these tests are not reliable is essentially the same as has already been discussed in Section 8.4 for a related test used with discriminant function analysis. This is that the ith largest sample canonical correlation may in fact have arisen from a population canonical correlation that is not the ith largest. Hence, the association between the r contributions to the right-hand side of Equation 10.3 and the r population canonical correlations is blurred. See Harris (2013) for a further discussion of this matter.

There are also some modifications of the test statistic X^2 that are sometimes proposed to improve the chi-squared approximation for the distribution of this statistic when the null hypothesis holds and the sample size is small, but these will not be considered here.

10.4   Interpreting canonical variates

If

$$U_i = a_{i1}X_1 + a_{i2}X_2 + \cdots + a_{ip}X_p$$

and

$$V_i = b_{i1}Y_1 + b_{i2}Y_2 + \cdots + b_{iq}Y_q$$

then it seems that U_i can be interpreted in terms of the X variables with large coefficients a_{ij}, and V_i can be interpreted in terms of the Y variables with large coefficients b_{ij}. Of course, *large* here means large positive or large negative.

Unfortunately, correlations between the X and Y variables can upset this interpretation process. For example, it can happen that a_{i1} is positive, and yet the simple correlation between U_i and X_1 is negative. This apparent contradiction can come about because X_1 is highly correlated with one or more of the other X variables, and part of the effect of X_1 is being accounted for by the coefficients of these other X variables. In fact, if one of the X variables is almost a linear combination of the other X variables, then there will be an infinite variety of linear combinations of the X variables, some of them with very different a_{ij} values, that give virtually the same U_1 values. The same can be said about linear combinations of the Y variables.

The interpretation problems that arise with highly correlated X or Y variables should be familiar to users of multiple regression analysis. Exactly the same problems arise with the estimation of regression coefficients.

Actually, a fair comment seems to be that if the X or Y variables are highly correlated, then there can be no way of disentangling their contributions to canonical variables. However, no doubt people will continue to try to make interpretations under these circumstances.

Some authors have suggested that it is better to describe canonical variables by looking at their correlations with the X and Y variables rather than the coefficients a_{ij} and b_{ij}. For example, if U_i is highly positively correlated with X_1, then U_i can be considered to reflect X_1 to a large extent. Similarly, if V_i is highly negatively correlated with Y_1, then V_i can be considered to reflect the opposite of Y_1 to a large extent. This approach does at least have the merit of bringing out all the variables to which the canonical variables seem to be related.

Example 10.1: Environmental and genetic correlations for colonies of a butterfly

The data in Table 1.3 can be used to illustrate the procedure for a canonical correlation analysis. Here, there are 16 colonies of the butterfly *Euphydryas editha* in California and Oregon. These vary with respect to four environmental variables (altitude, annual precipitation, annual maximum temperature, and annual minimum temperature) and six genetic variables (percentages of six phospho-glucose-isomerase genes as determined by electrophoresis). Any significant relationships between the environmental and genetic variables are interesting, because they may indicate the adaption of *E. editha* to local environments.

For a canonical correlation analysis, the environmental variables have been treated as the X variables and the gene frequencies as the Y variables. However, not all the six gene frequencies have been used, as shown in Table 1.3, because they add up to 100%, which allows different linear combinations of these variables to have the same correlation with a combination of the X variables. To see this, suppose that the first pair of canonical variables are U_1 and V_1, where

$$V_1 = b_{11}Y_1 + b_{12}Y_2 + \cdots + b_{16}Y_6$$

Then, V_1 can be rewritten by replacing Y_1 by 100 minus the sum of the other variables to give

$$V_1 = 100b_{11} + (b_{12} - b_{11})Y_2 + \cdots + (b_{16} - b_{11})Y_6$$

This means that the correlation between U_1 and V_1 is the same as that between

$$(b_{12} - b_{11})Y_2 + \cdots + (b_{16} - b_{11})Y_6$$

and U_1, because the constant $100b_{11}$ in the second linear combination has no effect on the correlation. Thus, two linear combinations of the Y variables, possibly with very different coefficients, can serve just as well for the canonical variable. In fact, it can be shown that an infinite number of different linear combinations of the Y variables will serve just as well, and the same holds true for linear combinations of standardized Y variables.

This problem is overcome by removing one of the gene frequencies from the analysis. Here, the 1.30 gene frequency was omitted. The data were also further modified by combining the low frequencies for the 0.40 and 0.60 mobility genes. Thus, the X variables being considered are X_1 = altitude, X_2 = annual precipitation, X_3 = annual maximum temperature, and X_4 = annual minimum temperature, while the Y variables are Y_1 = frequency of 0.40 and 0.60 mobility genes, Y_2 = frequency of 0.80 mobility genes, Y_3 = frequency of 1.00 mobility genes, and Y_4 = frequency of 1.16 mobility genes. Following the development in Section 10.2, it is the standardized values of the variables that have been analyzed, so that for the remainder of this example, X_i and Y_i, refer to the standardized X and Y variables.

Table 10.1 The correlation matrix for variables measured on colonies of *Euphydryas editha*, partitioned into **A**, **B**, **C**, and **C'** submatrices

	X_1	X_2	X_3	X_4	Y_1	Y_2	Y_3	Y_4
X_1	1.000	0.568	−0.828	−0.936	−0.201	−0.573	0.727	−0.458
X_2	0.568	1.000	−0.479	−0.705	−0.468	−0.550	0.699	−0.138
X_3	−0.828	−0.479	1.000	0.719	0.224	0.536	−0.717	0.438
X_4	−0.936	0.705	0.719	1.000	0.246	0.593	−0.759	0.412
					A	**C**		
					C'	**B**		
Y_1	−0.201	−0.468	0.224	0.246	1.000	0.638	−0.561	−0.584
Y_2	−0.573	−0.550	0.536	0.593	0.638	1.000	−0.824	−0.127
Y_3	0.727	0.699	−0.717	−0.759	−0.561	−0.824	1.000	−0.264
Y_4	−0.458	−0.138	0.438	0.412	−0.584	−0.127	−0.264	1.000

The correlation matrix for the eight variables is shown in Table 10.1, partitioned into the submatrices **A**, **B**, **C**, and **C'**, as described in Section 10.2. The eigenvalues obtained from Equation 10.1 are 0.7425, 0.2049, 0.1425, and 0.0069. Taking square roots gives the corresponding canonical correlations of 0.8617, 0.4527, 0.3775, and 0.0833, respectively, and the canonical variables are found to be

$$U_1 = -0.12X_1 - 0.29X_2 + 0.47X_3 + 0.26X_4$$
$$V_1 = +0.55Y_1 + 0.42Y_2 - 0.09Y_3 + 0.83Y_4$$
$$U_2 = +2.43X_1 - 0.68X_2 + 0.48X_3 + 1.40X_4$$
$$V_2 = -1.77Y_1 - 2.26Y_2 - 3.85Y_3 - 2.85Y_4$$
$$U_3 = +2.95X_1 + 1.36X_2 + 0.58X_3 + 3.53X_4$$
$$V_3 = -3.48Y_1 - 1.30Y_2 - 3.75Y_3 - 2.75Y_4$$

$$U_4 = +1.37X_1 + 0.24X_2 + 1.70X_3 - 0.09X_4$$

and

$$V_4 = +0.66Y_1 - 1.41Y_2 - 0.50Y_3 + 0.64Y_4$$

There are four canonical correlations, because this is the minimum of the number of X variables and the number of Y variables (which are both equal to four).

It should be noted that some statistical packages may give one or more of the equations with opposite signs, for example with $U_1 = 0.12X_1 + 0.29X_2 - 0.47X_3 - 0.26X_4$. This would reverse the meaning of U_1 but not its usefulness for describing the data.

Although the canonical correlations are quite large, they are not significantly so according to Bartlett's test, because of the small sample size. It is found that $X^2 = 18.34$ with 16 df, where the probability of a value this large from the chi-squared distribution is about 0.30.

Putting aside the lack of significance, it is interesting to see what interpretation can be given to the first pair of canonical variables. From the equation for U_1, it can be seen that this is mainly a contrast between X_3 (maximum temperature) and X_4 (minimum temperature), on the one hand, and X_2 (precipitation), on the other. For V_1, there are moderate to large positive coefficients for Y_1 (0.40 and 0.60 mobility), Y_2 (0.80 mobility), and Y_4 (1.16 mobility), and a small negative coefficient for Y_3 (1.00 mobility). It appears that the 0.40, 0.60, 0.80, and 1.16 mobility genes tend to be frequent in colonies with high temperatures and low precipitation.

The correlations between the environmental variables and U_1 are

	Altitude	Precipitation	Maximum temperature	Minimum temperature
U_1	−0.92	−0.77	0.90	0.92

suggesting that U_1 is best interpreted as a measure of high temperatures and low altitude and precipitation. The correlations between V_1 and the gene frequencies are

	Mobility 0.40/0.60	Mobility 0.80	Mobility 1.00	Mobility 1.16
V_1	0.38	0.74	−0.96	0.48

In this case, V_1 comes out clearly as indicating a lack of mobility 1.00 genes.

The interpretations of U_1 and V_1 are not the same when made on the basis of the coefficients of the canonical functions as they are on the basis of correlations. For U_1, the difference is not great and only concerns the status of altitude, but for V_1, the importance of the mobility 1.00 genes is very different. On the whole, the interpretations based on correlations seem best and correspond with what is seen in the data. For example, colony GL has the highest altitude, high precipitation, the lowest temperatures, and the highest frequency of 1.00 mobility genes. This compares with colony UO with a low altitude, low precipitation, high temperature, and the lowest frequency of mobility 1.00 genes. However, as mentioned in the previous section, there are real problems with interpreting canonical variables when the variables that they are constructed from have high correlations. Table 10.1 shows that this is indeed the case with this example.

Figure 10.1 shows a plot of the values of V_1 against the values of U_1. It is immediately clear that the colony labeled DP is somewhat

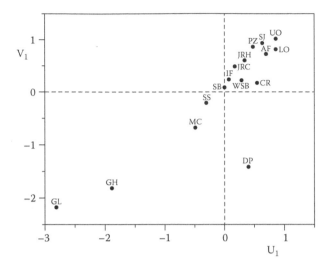

Figure 10.1 Plot of V_1 against U_1 for 16 colonies of *Euphydryas editha*.

unusual compared with the other colonies, because the value of V_1 is not similar to that for other colonies with about the same values for U_1. From the interpretations given for U_1 and V_1, it would seem that the frequency of mobility 1.00 genes is unusually high for a colony with this environment. Inspection of the data in Table 1.3 shows that this is the case.

Example 10.2: Soil and vegetation variables in Belize

For an example with a larger data set, consider part of the data collected by Green (1973) for a study on the factors influencing the location of prehistoric Maya habitation sites in the Corozal district of Belize in Central America. Table 10.2 shows four soil variables and four vegetation variables recorded for 2.5 × 2.5 km squares. Here, canonical correlation analysis can be used to study the relationship between these two groups of variables.

The soil variables are $X_1 = \%$ soil with constant lime enrichment, $X_2 = \%$ meadow soil with calcium groundwater, $X_3 = \%$ soil with coral bedrock under conditions of constant lime enrichment, and $X_4 = \%$ alluvial and organic soils adjacent to rivers and saline organic soil at the coast. The vegetation variables are $Y_1 = \%$ deciduous seasonal broadleaf forest, $Y_2 = \%$ high and low marsh forest, herbaceous marsh, and swamp, $Y_3 = \%$ cohune palm forest, and $Y_4 = \%$ mixed forest. The percentages do not add to 100, for all the squares, so there is no need to remove any variables before starting the analysis. It is the standardized values of these variables, with means of zero and standard deviations of one, that will be referred to for the rest of this example.

Table 10.2 Soil and vegetation variables for 151 squares, 2.5 × 2.5 km, in the Corozal region of Belize

Square	X_1	X_2	X_3	X_4	Y_1	Y_2	Y_3	Y_4
1	40	30	0	30	0	25	0	0
2	20	0	0	10	10	90	0	0
3	5	0	0	50	20	50	0	0
4	30	0	0	30	0	60	0	0
5	40	20	0	20	0	95	0	0
6	60	0	0	5	0	100	0	0
7	90	0	0	10	0	100	0	0
8	100	0	0	0	20	80	0	0
9	0	0	0	10	40	60	0	0
10	15	0	0	20	25	10	0	0
11	20	0	0	10	5	50	0	0
12	0	0	0	50	5	60	0	0
13	10	0	0	30	30	60	0	0
14	40	0	0	20	50	10	0	0
15	10	0	0	40	80	20	0	0
16	60	0	0	0	100	0	0	0
17	45	0	0	0	5	60	0	0
18	100	0	0	0	100	0	0	0
19	20	0	0	0	20	0	0	0
20	0	0	0	60	0	50	0	0
21	0	0	0	80	0	75	0	0
22	0	0	0	50	0	50	0	0
23	30	10	0	60	0	100	0	0
24	0	0	0	50	0	50	0	0
25	50	20	0	30	0	100	0	0
26	5	15	0	80	0	100	0	0
27	60	40	0	0	10	90	0	0
28	60	40	0	0	50	50	0	0
29	94	5	0	0	90	10	0	0
30	80	0	0	20	0	100	0	0
31	50	50	0	0	25	75	0	0
32	10	40	50	0	75	25	0	0
33	12	12	75	0	10	90	0	0
34	50	50	0	0	15	85	0	0
35	50	40	10	0	80	20	0	0
36	0	0	100	0	100	0	0	0
37	0	0	100	0	100	0	0	0
38	70	30	0	0	50	50	0	0

(Continued)

Table 10.2 (Continued) Soil and vegetation variables for 151
squares, 2.5 × 2.5 km, in the Corozal region of Belize

Square	X_1	X_2	X_3	X_4	Y_1	Y_2	Y_3	Y_4
39	40	40	20	0	50	50	0	0
40	0	0	100	0	100	0	0	0
41	25	25	50	0	100	0	0	0
42	40	40	0	20	80	20	0	0
43	90	0	0	10	100	0	0	0
44	100	0	0	0	100	0	0	0
45	100	0	0	0	90	10	0	0
46	10	0	0	90	100	0	0	0
47	80	0	0	20	100	0	0	0
48	60	0	0	30	80	0	0	0
49	40	0	0	0	0	30	0	0
50	50	0	0	50	100	0	0	0
51	50	0	0	0	40	0	0	0
52	30	30	0	20	30	60	0	0
53	20	20	0	40	0	100	0	0
54	20	80	0	0	0	100	0	0
55	0	10	0	60	0	75	0	0
56	0	50	0	30	0	75	0	0
57	50	50	0	0	30	70	0	0
58	0	0	0	60	0	60	0	0
59	20	20	0	60	0	100	0	0
60	90	10	0	0	70	30	0	0
61	100	0	0	0	100	0	0	0
62	15	15	0	30	0	40	0	0
63	100	0	0	0	25	75	0	0
64	95	0	0	5	90	10	0	0
65	95	0	0	5	90	10	0	0
66	60	40	0	0	50	50	0	0
67	30	60	10	10	50	10	0	0
68	50	0	50	50	100	0	0	0
69	60	30	0	10	60	40	0	0
70	90	8	0	2	80	20	0	0
71	30	30	30	40	60	40	0	0
72	33	33	33	33	75	25	0	0
73	20	10	0	40	0	100	0	0
74	50	0	0	50	40	60	0	0
75	75	12	0	12	50	50	0	0
76	75	0	0	25	40	60	0	0

(Continued)

Table 10.2 (Continued) Soil and vegetation variables for 151 squares, 2.5 × 2.5 km, in the Corozal region of Belize

Square	X_1	X_2	X_3	X_4	Y_1	Y_2	Y_3	Y_4
77	30	0	0	50	0	100	0	0
78	50	10	0	30	5	95	0	0
79	100	0	0	0	60	40	0	0
80	50	0	0	50	20	80	0	0
81	10	0	0	90	0	100	0	0
82	30	30	0	20	0	85	0	0
83	20	20	0	20	0	75	0	0
84	90	0	0	0	50	25	0	0
85	30	0	0	0	30	5	0	0
86	20	30	0	50	20	80	0	0
87	50	30	0	10	50	50	0	0
88	80	0	0	0	70	10	0	0
89	80	0	0	0	50	0	0	0
90	60	10	0	25	80	15	0	0
91	50	0	0	0	75	0	0	0
92	70	0	0	0	75	0	0	0
93	100	0	0	0	85	15	0	0
94	60	30	0	0	40	60	0	0
95	80	20	0	0	50	50	0	0
96	100	0	0	0	100	0	0	0
97	100	0	0	0	95	5	0	0
98	0	0	0	60	0	50	0	0
99	30	20	0	30	0	60	0	40
100	15	0	0	35	20	30	0	0
101	40	0	0	45	70	20	0	0
102	30	0	0	45	20	40	0	20
103	60	10	0	30	10	65	5	20
104	40	20	0	40	0	25	0	75
105	100	0	0	0	70	0	0	30
106	100	0	0	0	40	60	0	0
107	80	10	0	10	40	60	0	0
108	90	0	0	10	10	0	0	90
109	100	0	0	0	20	10	0	70
110	30	50	0	20	10	90	0	0
111	60	40	0	0	50	50	0	0
112	100	0	0	0	80	10	0	10
113	60	0	0	40	60	10	30	0
114	50	50	0	0	0	100	0	0

(Continued)

Table 10.2 (Continued) Soil and vegetation variables for 151 squares, 2.5 × 2.5 km, in the Corozal region of Belize

Square	X_1	X_2	X_3	X_4	Y_1	Y_2	Y_3	Y_4
115	60	30	0	10	25	75	0	0
116	40	0	0	60	30	20	50	0
117	30	0	0	70	0	50	50	0
118	50	20	0	30	0	100	0	0
119	50	50	0	0	25	75	0	0
120	90	10	0	0	50	50	0	0
121	100	0	0	0	60	40	0	0
122	50	0	0	50	70	30	0	0
123	10	10	0	80	0	100	0	0
124	50	50	0	0	30	70	0	0
125	75	0	0	25	80	20	0	0
126	40	0	0	60	0	100	0	0
127	90	10	0	10	75	25	0	0
128	45	45	0	55	30	70	0	0
129	20	35	0	80	10	90	0	0
130	80	0	0	20	70	30	0	0
131	100	0	0	0	90	0	0	0
132	75	0	0	25	50	50	0	0
133	60	5	0	40	50	50	0	0
134	40	0	0	60	60	40	0	0
135	60	0	0	40	70	15	0	0
136	90	10	0	10	75	25	0	0
137	50	0	5	0	30	20	0	0
138	70	0	30	0	70	30	0	0
139	60	0	40	0	100	0	0	0
140	50	0	0	0	50	0	0	0
141	30	0	50	0	60	40	0	0
142	5	0	95	0	80	20	0	0
143	10	0	90	0	70	30	0	0
144	50	0	0	0	15	30	0	0
145	20	0	80	0	50	50	0	0
146	0	0	100	0	90	10	0	0
147	0	0	100	0	75	25	0	0
148	90	0	10	0	60	30	10	0
149	0	0	100	0	80	10	10	0
150	0	0	100	0	60	40	0	0
151	0	40	60	40	50	50	0	0

Note: X_1 = % soil with constant lime enrichment, X_2 = % meadow soil with calcium groundwater, X_3 = % soil with coral bedrock under conditions of constant lime enrichment, and X_4 = % alluvial and organic soils adjacent to rivers and saline organic soil at the coast. Y_1 = % deciduous seasonal broadleaf forest, Y_2 = % high and low marsh forest, herbaceous marsh, and swamp, Y_3 = % cohune palm forest, and Y_4 = % mixed forest.

There are four canonical correlations (the minimum of the number of X variables and the number of Y variables), and they are found to be 0.762, 0.566, 0.243, and 0.122. The X^2 statistic of Equation 10.3 is found to equal 193.63 with 16 df, which is very highly significantly large when compared with the percentage points of the chi-squared distribution. Therefore, there is very strong evidence that the soil and vegetation variables are related. However, the original data are clearly not normally distributed, so this result should be treated with some reservations.

The canonical variables are found to be

$$U_1 = +1.33X_1 + 0.29X_2 + 1.12X_3 + 0.56X_4$$
$$V_1 = +1.67Y_1 + 1.00Y_2 + 0.21Y_3 + 0.51Y_4$$
$$U_2 = +0.49X_1 + 0.88X_2 + 0.29X_3 + 0.97X_4$$
$$V_2 = +0.70Y_1 + 1.50Y_2 + 0.32Y_3 + 0.31Y_4$$
$$U_3 = +0.38X_1 - 0.57X_2 + 0.14X_3 + 0.87X_4$$
$$V_3 = -0.22Y_1 - 0.30Y_2 + 0.92Y_3 + 0.20Y_4$$
$$U_4 = -0.44X_1 - 0.02X_2 + 0.72X_3 + 0.15X_4$$

and

$$V_4 = +0.12Y_1 + 0.01Y_2 + 0.26Y_3 - 0.93Y_4$$

In fact, the linear combinations given here for U_1, V_1, U_2, and V_2 are not the ones that were output by the program used to do the calculations, because the output linear combinations all had negative coefficients for the X and Y variables. A switch in sign is justified, because the correlation between $-U_i$ and $-V_i$ is the same as that between U_i and V_i. Hence, $-U_i$ and $-V_i$ will serve as well as U_i and V_i as the ith canonical variables. Note, however, that switching signs for U_1, V_1, U_2, and V_2 has changed the signs of the correlations between these canonical variables and the X and Y variables, as shown in Table 10.3.

By considering the correlations shown in Table 10.3 (particularly those outside the range from −0.5 to +0.5), it appears that the canonical variables can be described as mainly measuring

Table 10.3 Correlations between the canonical variables and the X and Y variables

	U_1	U_2	U_3	U_4		V_1	V_2	V_3	V_4
X_1	0.57	−0.20	0.00	−0.80	Y_1	0.80	−0.55	−0.07	0.24
X_2	−0.06	0.69	−0.72	−0.04	Y_2	−0.40	0.89	−0.22	0.03
X_3	0.42	−0.23	−0.17	0.86	Y_3	0.04	0.17	0.95	0.28
X_4	−0.36	0.58	0.71	0.19	Y_4	0.11	−0.01	0.25	−0.96

U₁: the presence of soil types 1 (soil with constant lime enrichment) and 3 (soil with coral bedrock under conditions of constant lime enrichment)

V₁: the presence of vegetation type 1 (deciduous seasonal broadleaf forest)

U₂: the presence of soil types 2 (meadow soil with calcium groundwater) and 4 (alluvial and organic soils adjacent to rivers and saline organic soil at the coast)

V₂: the presence of vegetation type 2 (high and low marsh forest, herbaceous marsh, and swamp) and the absence of vegetation type 1

U₃: the presence of soil type 4 and the absence of soil type 2

V₃: the presence of vegetation type 3 (cohune palm forest)

U₄: the presence of soil type 3 and the absence of soil type 1

V₄: the absence of vegetation type 4 (mixed forest).

It appears, therefore, that the most important relationships between the soil and vegetation variables, as described by the first two pairs of canonical variables, are (a) the presence of soil types 1 and 3 and the absence of soil type 4 are associated with the presence of vegetation type 1, and (b) the presence of soil types 2 and 4 is associated with the presence of vegetation type 2 and the absence of vegetation type 1.

It is instructive to examine a plot of the canonical variables and the case numbers, as shown in Figure 10.2. The correlations between U_1 and V_1 and between U_2 and V_2 are apparent, as might be expected. Perhaps the most striking thing shown by the plots is the unusual distributions of V_3 and V_4. Most values are very similar, at about -0.2

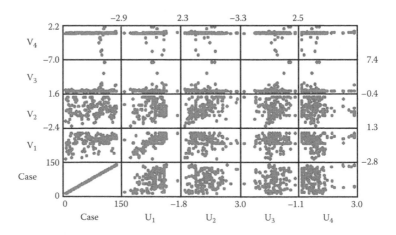

Figure 10.2 Draftsman's plot of canonical variables obtained from the data on soil and vegetation variables for 2.5 km squares in Belize. (Note that to enhance readability, some of the unit scales for the U and V axes appear, respectively, above and to the right of the plots.)

for V_3 and about +0.2 for V_4. However, there are extreme values for some cases (observations) between 100 and 120. Inspection of the data in Table 10.2 shows that these extreme cases are for squares where vegetation types 3 and 4 were present, which makes perfect sense from the definitions of V_3 and V_4.

Before leaving this example, it is appropriate to mention a potential problem that has not been addressed. This concerns the spatial correlation in the data for squares that are close in space, and particularly those that are adjacent. If such correlation exists, so that, for example, neighboring squares tend to have the same soil and vegetation characteristics, then the data do not provide 151 independent observations. In effect, the data set will be equivalent to independent data from some smaller number of squares. The effect of this will appear mainly in the test for the significance of the canonical correlations as a whole, with a tendency for these correlations to appear to be more significant than they really are.

The same problem also potentially exists with the previous example on colonies of the butterfly *Euphydryas editha*, because some of the colonies were quite close in space. Indeed, it is a potential problem whenever observations are taken in different locations in space. The way to avoid the problem is to ensure that observations are taken sufficiently far apart that they are independent or close to independent, although this is often easier said than done. There are methods available for allowing for spatial correlation in data, but these are beyond the scope of this book.

10.5   Computer programs

The Appendix to this chapter provides information about R packages that can be used to carry out the analyses described in this chapter. However, the option for canonical correlation analysis is not as widely available in statistical packages as the options for the multivariate analyses considered in earlier chapters. Still, the larger packages certainly provide it.

10.6   Further reading

There are not many books available that concentrate only on the theory and applications of canonical correlation analysis. A useful reference is the book by Giffins (1985) on applications of canonical correlation analysis in ecology. About half of this text is devoted to theory and the remainder to specific examples on plants. A shorter text with a social sciences emphasis is by Thompson (1985).

Exercise

Table 10.4 shows the result of combining the data in Tables 1.5 and 6.7 on sources of protein and employment patterns for European countries, for

Table 10.4 Sources of protein and percentages employed in different industry groups for countries in Europe

Country	RM	WM	EGG	MLK	FSH	CRL	SCH	PNO	F&V	AGR	MIN	MAN	PS	CON	SER	FIN	SPS	TC
Albania	10	1	1	9	0	42	1	6	2	55.5	19.4	0.0	0.0	3.4	3.3	15.3	0.0	3.0
Austria	9	14	4	20	2	28	4	1	4	7.4	0.3	26.9	1.2	8.5	19.1	6.7	23.3	6.4
Belgium	14	9	4	18	5	27	6	2	4	2.6	0.2	20.8	0.8	6.3	16.9	8.7	36.9	6.8
Bulgaria	8	6	2	8	1	57	1	4	4	19.0	0.0	35.0	0.0	6.7	9.4	1.5	20.9	7.5
Denmark	11	11	4	25	10	22	5	1	2	5.6	0.1	20.4	0.7	6.4	14.5	9.1	36.3	7.0
Finland	10	5	3	34	6	26	5	1	1	8.5	0.2	19.3	1.2	6.8	14.6	8.6	33.2	7.5
France	18	10	3	20	6	28	5	2	7	5.1	0.3	20.2	0.9	7.1	16.7	10.2	33.1	6.4
Greece	10	3	3	18	6	42	2	8	7	22.2	0.5	19.2	1.0	6.8	18.2	5.3	19.8	6.9
Hungary	5	12	3	10	0	40	4	5	4	15.3	28.9	0.0	0.0	6.4	13.3	0.0	27.3	8.8
Ireland	14	10	5	26	2	24	6	2	3	13.8	0.6	19.8	1.2	7.1	17.8	8.4	25.5	5.8
Italy	9	5	3	14	3	37	2	4	7	8.4	1.1	21.9	0.0	9.1	21.6	4.6	28.0	5.3
The Netherlands	10	14	4	23	3	22	4	2	4	4.2	0.1	19.2	0.7	0.6	18.5	11.5	38.3	6.8
Norway	9	5	3	23	10	23	5	2	3	5.8	1.1	14.6	1.1	6.5	17.6	7.6	37.5	8.1
Poland	7	10	3	19	3	36	6	2	7	23.6	3.9	24.1	0.9	6.3	10.3	1.3	24.5	5.2
Portugal	6	4	1	5	14	27	6	5	8	11.5	0.5	23.6	0.7	8.2	19.8	6.3	24.6	4.8
Romania	6	6	2	11	1	50	3	5	3	22.0	2.6	37.9	2.0	5.8	6.9	0.6	15.3	6.8
Spain	7	3	3	9	7	29	6	6	7	9.9	0.5	21.1	0.6	9.5	20.1	5.9	26.7	5.8
Sweden	10	8	4	25	8	20	4	1	2	3.2	0.3	19.0	0.8	6.4	14.2	9.4	39.5	7.2
Switzerland	13	10	3	24	2	26	3	2	5	5.6	0.0	24.7	0.0	9.2	20.5	10.7	23.1	6.2
United Kingdom	17	6	5	21	4	24	5	3	3	2.2	0.7	21.3	1.2	7.0	20.2	12.4	28.4	6.5
USSR (former)	9	5	2	17	3	44	6	3	3	18.5	0.0	28.8	0.0	10.2	7.9	0.6	25.6	8.4
Yugoslavia (former)	4	5	1	10	1	56	3	6	3	5.0	2.2	38.7	2.2	8.1	13.8	3.1	19.1	7.8

Note: RM, red meat; WM, white meat; EGG, eggs; MLK, milk; FSH, fish; CLR, cereals; SCH, starchy foods; PNO, pulses, nuts, and oilseed; F&V, fruit and vegetables; AGR, agriculture, forestry, and fishing; MIN, mining and quarrying; MAN, manufacturing; PS, power and water supplies; CON, construction; SER, services; FIN, finance; SPS, social and personal services; TC, transport and communications.

the 22 countries where this is possible. Use canonical correlation analysis to investigate the relationship, if any, between the nature of the employment in a country and the type of food that is used for protein.

References

Bartlett, M.S. (1947). The general canonical correlation distribution. *Annals of Mathematical Statistics* 18: 1–17.

Giffins, R. (1985). *Canonical Analysis: A Review with Applications in Ecology.* Berlin: Springer.

Green, E.L. (1973). Location analysis of prehistoric Maya sites in British Honduras. *American Antiquity* 38: 279–93.

Harris, R.J. (2013). *A Primer of Multivariate Statistics.* 3rd Edn. New York and Hove: Psychology Press.

Hotelling, H. (1936). Relations between two sets of variables. *Biometrika* 28: 321–77.

Thompson, B. (1985). *Canonical Correlation Analysis: Uses and Interpretations.* Thousand Oaks, CA: Sage.

Appendix: Canonical Correlation in R

Canonical correlation analysis is basically an eigenvalue problem, as defined by Equation 10.1, so it becomes evident that the canonical correlations for two sets of variables (say, X and Y variables) are easily computed in R with the function `eigen()`. The programming effort is not remarkably reduced if the user decides to execute `cancor(matX, matY)`, the basic command offered by R in the default package `stats` for canonical correlation analysis. Here, `matX` and `matY` are matrices containing the X and Y variables, respectively. `Cancor()` only allows data-centering, so `scale()` must be run each time data standardization is required. The output of `cancor()` is quite simple. It consists of the correlations, the coefficients of the linear combinations defining the canonical variables, and the means of the X and Y variables included in the analysis.

A wrapper of `cancor()` (i.e. a function with the same name) has been included in `candisc` (Friendly and Fox, 2016), a package that was already considered in the Appendix to Chapter 8 for discriminant function analysis. The function `cancor()` implemented in `candisc` allows more general calculations and plots of the canonical variables in two dimensions. It also permits data standardization and provides a set of Wilks' lambda tests as alternatives to the Bartlett's tests described in Section 10.3. Moreover, the correlation matrices between the X and Y variables, and their corresponding canonical variables U_i and V_i, are part of the output produced by this improved version of `cancor()`.

Bartlett's chi-squared test is included as the function `cca()` found in `yacca` (Butts, 2012), a package whose acronym stands for "yet another package for canonical correlation analysis." The name `cca` is unfortunate here, as the data analyst may be confused by the same name being commonly used for the multivariate method *constrained correspondence analysis* (Oksanen, 2016). This method is mentioned in Chapter 12 with the name *canonical correspondence analysis* (Legendre and Legendre, 2012). A typical command involving `cca` for canonical correlation analysis takes the form

```
cca.object <- cca(matX, matY, xscale = TRUE, yscale = TRUE)
```

Here, `matX` and `matY` are the unstandardized data matrices whose columns are internally standardized by `cca` as a response to the options `xscale=TRUE` and `yscale=TRUE`.

There are two more packages containing functions for canonical correlation analysis in R. One is the CCA package (González and Déjean, 2012), which includes the function `cc()`, a regularized version of canonical correlation analysis to deal with data sets with more variables than units. The second package is `vegan` (Oksanen et al., 2016), in which a function called CCorA() permits better canonical correlation analyses in cases of very sparse (with many zeros) and collinear matrices (with linearly dependent columns). The reader is encouraged to review the details of these two functions in the references given below and the R help documentation.

We recommend cancor(), implemented in candisc, and the function cca from yacca as the most suitable functions for canonical correlation analysis in R. In fact, both commands are sufficient to produce the results for Examples 10.1 and 10.2. The corresponding R scripts are available from the book's website.

References

Butts, C.T. (2012). Yacca: Yet Another Canonical Correlation Analysis Package. R package version 1.1. https://CRAN.R-project.org/package=yacca

Friendly, M. and Fox, J. (2016). candisc: Visualizing Generalized Canonical Discriminant and Canonical Correlation Analysis. R package version 0.7-0. http://CRAN.R-project.org/package=candisc

González, I. and Déjean, S. (2012). CCA: Canonical correlation analysis. R package version 1.2. https://CRAN.R-project.org/package=CCA

Legendre, P. and Legendre, L. (2012). *Numerical Ecology*. 3rd Edn. Amsterdam: Elsevier.

Oksanen, J. (2016). *Vegan: An Introduction to Ordination*. https://cran.r-project.org/web/packages/vegan/vignettes/intro-vegan.pdf

Oksanen, J., Blanchet, F.G., Friendly, M., Kindt, R., Legendre, P., McGlinn, D., Minchin, P.R., et al. (2016). vegan: Community Ecology Package. R package version 2.4-0. http://CRAN.R-project.org/package=vegan

chapter eleven

Multidimensional scaling

11.1   Constructing a map from a distance matrix

Multidimensional scaling is designed to construct a diagram showing the relationships between a number of objects, given only a table of distances between the objects. The diagram is then a type of map, which can be in one dimension (if the objects fall on a line), in two dimensions (if the objects lie on a plane), in three dimensions (if the objects can be represented by points in space), or in a higher number of dimensions (in which case a simple geometrical representation is not possible).

The fact that it may be possible to construct a map from a table of distances can be seen by considering the example of four objects A, B, C, and D shown in Figure 11.1. The distances between the objects are given in Table 11.1. For example, the distance from A to B, which is the same as the distance from B to A, is 6.0, while the distance from each object to itself is always 0.0. It seems plausible that the map can be reconstructed from the array of distances. However, it is also apparent that a mirror image of the map, as shown in Figure 11.2, will have the same array of distances between objects. Consequently, it seems clear that a recovery of the original map will be subject to a possible reversal of this type.

It is also apparent that if more than three objects are involved, then they may not lie on a plane. In that case, the distance matrix will implicitly contain this information. For example, the distance array shown in Table 11.2 requires three dimensions to show the spatial relationships between the four objects. Unfortunately, with real data, it is not usually known how many dimensions are needed for a representation. Hence, with real data, a range of dimensions usually has to be tried.

The usefulness of multidimensional scaling comes from the fact that situations often arise where the underlying relationship between objects is not known, but a distance matrix can be estimated. For example, in psychology, subjects may be able to assess how similar or different individual pairs of objects are without being able to draw an overall picture of the relationships between the objects. Multidimensional scaling can then provide the picture.

At the present time, there are a wide variety of data analysis techniques that go under the general heading of multidimensional scaling. Here, only the simplest of these will be considered, these being the classical methods

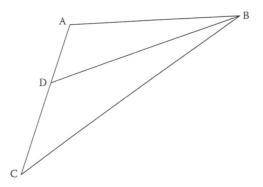

Figure 11.1 Four objects in two dimensions.

Table 11.1 Euclidean distances between the
objects shown in Figure 11.1

	A	B	C	D
A	0.0	6.0	6.0	2.5
B	6.0	0.0	9.5	7.8
C	6.0	9.5	0.0	3.5
D	2.5	7.8	3.5	0.0

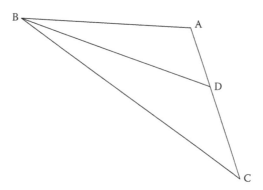

Figure 11.2 A mirror image of the objects in Figure 11.1, for which the distances
between the objects are the same.

proposed by Togerson (1952) and Kruskal (1964a,b). A related method
called *principal coordinates analysis* is discussed in Chapter 12.

11.2   Procedure for multidimensional scaling

Classical multidimensional scaling starts with a matrix of distances
between n objects that has δ_{ij}, the distance from object i to object j, in the

Table 11.2 A matrix of distances between
four objects in three dimensions

	A	B	C	D
A	0	1	$\sqrt{2}$	$\sqrt{2}$
B	1	0	1	1
C	$\sqrt{2}$	1	0	$\sqrt{2}$
D	$\sqrt{2}$	1	$\sqrt{2}$	0

ith row and the jth column. The number of dimensions for the mapping of objects is fixed for a particular solution at t (1 or more). Different computer programs use different methods for carrying out analysis, but generally something like the following steps is involved.

1. A starting configuration is set up for the n objects in t dimensions, that is, coordinates $(x_1, x_2, \ldots , x_t)$ are assumed for each object in a t-dimensional space.
2. The Euclidean distances between the objects are calculated for the assumed configuration. Let d_{ij} be the distance between object i and object j for this configuration.
3. A regression of d_{ij} on δ_{ij} is made, where, as mentioned at the beginning of this section, δ_{ij} is the distance between object i and j according to the input data. The regression can be linear, polynomial, or monotonic. For example, a linear regression assumes that

$$d_{ij} = \alpha + \beta\delta_{ij} + \varepsilon_{ij}$$

where:

   ε_{ij} is an error term
   α and β are constants

A monotonic regression just assumes that if δ_{ij} increases, then d_{ij} either increases or remains constant, but no exact relationship between δ_{ij} and d_{ij} is assumed. The fitted distances obtained from the regression equation ($\hat{d}_{ij} = \alpha + \beta\delta_{ij}$, assuming a linear regression) are called *disparities*. That is to say, the disparities $\hat{d}_{ij}$ are the data distances δ_{ij} scaled to match the configuration distance d_{ij} as closely as possible.

4. The goodness of fit between the configuration distances and the disparities is measured by a suitable statistic. One possibility is Kruskal's stress formula 1:

$$\text{STRESS1} = \left\{ \sum \left(d_{ij} - \hat{d}_{ij} \right)^2 / \sum \hat{d}_{ij}^2 \right\}^{1/2} \tag{11.1}$$

The word *stress* is used here because the statistic is a measure of the extent to which the spatial configuration of points has to be stressed in order to obtain the data distances δ_{ij}.

5. The coordinates $(x_1, x_2, \ldots, x_t)$ of each object are changed slightly in such a way that the stress is reduced.

Steps 2 to 5 are repeated until it seems that the stress cannot be further reduced. The outcome of the analysis is then the coordinates of the n objects in t dimensions. These coordinates can be used to draw a map that shows how the objects are related. It is desirable that a good solution is found in three or fewer dimensions, as a graphical representation of the n objects is then straightforward. Obviously, this is not always possible.

Small values of STRESS 1 (close to 0) are desirable. However, defining what is meant by "small" for a good solution is not straightforward. As a rough guide, Kruskal and Wish (1978, p. 56) indicate that reducing the number of dimensions to the extent that STRESS 1 exceeds 0.1, or increasing the number of dimensions when STRESS 1 is already less than 0.05, is questionable. However, their discussion concerning choosing the number of dimensions involves more considerations than this. In practice, the choice of the number of dimensions is often made subjectively, based on a compromise between the desire to keep the number small and the opposing desire to make the stress as small as possible. What is clear is that in general, there is little point in increasing the number of dimensions if this only leads to a small decrease in the stress.

An important distinction is between metric multidimensional scaling and nonmetric dimensional scaling. In the metric case, the configuration distances d_{ij} and the data distances δ_{ij} are related by a linear or polynomial regression equation. With nonmetric scaling, all that is required is a monotonic regression, which means that only the ordering of the data distances is important. Generally, the greater flexibility of nonmetric scaling should enable a better low-dimensional representation of the data to be obtained.

Example 11.1: Road distances between New Zealand towns

As an example of what can be achieved by multidimensional scaling, a map of the South Island of New Zealand has been constructed from a table of the road distances between the 13 towns shown in Figure 11.3.

If road distances were proportional to geographic distances, it would be possible to recover the true map exactly by a two-dimensional analysis. However, due to the absence of direct road links between many towns, road distances are in some cases far greater than geographic distances. Consequently, all that can be hoped for is a rather approximate recovery of the true map shown in Figure 11.3 from the road distances that are shown in Table 11.3.

Figure 11.3 The South Island of New Zealand, with the main roads between 13 towns indicated by broken lines.

The computer program NCSS (Hintze, 2012) was used for the analysis. At Step 3 of the procedure described on page 205, a monotonic regression relationship was assumed between the map distances d_{ij} and the distances δ_{ij} given in Table 11.3. This gives what is sometimes called a *classical nonmetric multidimensional scaling*.

The program produced a two-dimensional solution for the data using the algorithm described above. The final stress value was 0.041 as calculated using Equation 11.1.

The output from the program includes the coordinates of the 13 towns for the two dimensions produced in the analysis, as shown in Table 11.4. To maintain the north–south and east–west orientation

Table 11.3 Main road distances in miles between 13 towns in the South Island of New Zealand

	Alexandra	Balclutha	Blenheim	Christchurch	Dunedin	Franz Josef	Greymouth	Invercargill	Milford	Nelson	Queenstown	Te Anau	Timaru
Alexandra	—												
Balclutha	100	—											
Blenheim	485	478	—										
Christchurch	284	276	201	—									
Dunedin	126	50	427	226	—								
Franz Josef	233	493	327	247	354	—							
Greymouth	347	402	214	158	352	114	—						
Invercargill	138	89	567	365	139	380	493	—					
Milford	248	213	691	489	263	416	555	174	—				
Nelson	563	537	73	267	493	300	187	632	756	—			
Queenstown	56	156	494	305	192	228	341	118	178	572	—		
Te Anau	173	138	615	414	188	366	480	99	75	681	117	—	
Timaru	197	177	300	99	127	313	225	266	377	366	230	315	—

Table 11.4 Coordinates produced by multidimensional scaling applied to the distances between towns in the South Island of New Zealand

| Town | Dimension | | |
	1	2	New 2
Alexandra	0.11	0.07	−0.07
Balcluha	0.19	−0.08	0.08
Blenheim	−0.38	−0.16	0.16
Christchurch	−0.15	−0.11	0.11
Dunedin	0.13	−0.10	0.10
Franz Josef	−0.18	0.20	−0.20
Greymouth	−0.27	0.06	−0.06
Invercargill	0.26	−0.01	0.01
Milford	0.36	0.13	−0.13
Nelson	−0.45	−0.08	0.08
Queenstown	0.13	0.12	−0.12
Te Anau	0.28	0.08	−0.08
Timaru	−0.03	−0.13	0.13

Note: Dimension 2 is what was produced by the computer program used. The signs of this axis have been reversed for the new dimension 2 to match the geographical locations of the real towns.

that exists between the real towns, the signs of the values for the second dimension have been reversed to produce what is called *new dimension 2*. This sign reversal does not change the distances between the towns based on the two dimensions, and the new dimension is therefore just as satisfactory as the original one. If the sign is left unchanged, then the plot of the towns against the two dimensions looks like a mirror image of the real map.

A plot of the towns using these coordinates is shown in Figure 11.4. A comparison of this figure with Figure 11.3 indicates that the multidimensional scaling has been quite successful in recovering the real map. On the whole, the towns are shown with the correct relationships to each other. An exception is Milford. Because this town can only be reached by road through Te Anau, the map produced by multidimensional scaling has made Milford closest to Te Anau. In fact, Milford is geographically closer to Queenstown than it is to Te Anau.

Example 11.2: The voting behavior of congressmen

For a second example of the value of multidimensional scaling, consider the distance matrix shown in Table 11.5. Here, the

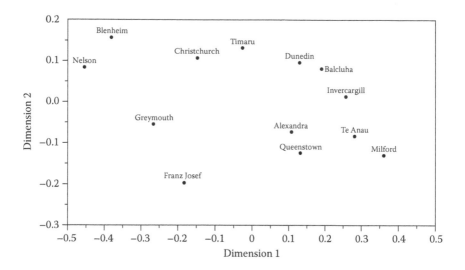

Figure 11.4 The map produced by a multidimensional scaling using the distances between New Zealand towns shown in Table 11.3.

distances are between 15 New Jersey congressmen in the United States House of Representatives. They are counts of the number of voting disagreements on 19 bills concerned with environmental matters. For example, Congressmen Hunt and Sandman disagreed 8 out of the 19 times, Sandman and Howard disagreed 17 out of the 19 times, and so on. An agreement was considered to occur if two congressmen both voted yes, both voted no, or both failed to vote. The table of distances was constructed from original data given by Romesburg (2004).

Two analyses were carried out using the NCSS (Hintze, 2012) program. The first was a classical metric multidimensional scaling, which assumes that the distances of Table 11.5 are measured on a ratio scale. That is to say, it is assumed that doubling a distance value is equivalent to assuming that the configuration distance between two objects is doubled. This means that the regression at Step 3 of the procedure described at the beginning of this section is of the form

$$d_{ij} = \beta \delta_{ij} + \varepsilon_{ij}$$

where:

ε_{ij} is an error term
β is a constant

The stress values obtained for two-, three-, and four-dimensional solutions were found on this basis to be 0.237, 0.130, and 0.081, respectively.

Table 11.5 The distances between 15 congressmen from New Jersey in the United States House of Representatives

	Hunt	Sandman	Howard	Thompson	Frelinghuysen	Forsythe	Widnall	Roe	Helstoski	Rodino	Minish	Rinaldo	Maraziti	Daniels	Pattern
Hunt (R)	0														
Sandman (R)	8	0													
Howard (D)	15	17	0												
Thompson (D)	15	12	9	0											
Frelinghuysen (R)	10	13	16	14	0										
Forsythe (R)	9	13	12	12	8	0									
Widnall (R)	7	12	15	13	9	7	0								
Roe (D)	15	16	5	10	13	12	17	0							
Helstoski (D)	16	17	5	8	14	11	16	4	0						
Rodino (D)	14	15	6	8	12	10	15	5	3	0					
Minish (D)	15	16	5	8	12	9	14	5	2	1	0				
Rinaldo (R)	16	17	4	6	12	10	15	3	1	2	1	0			
Maraziti (R)	7	13	11	15	10	6	10	12	13	11	12	12	0		
Daniels (D)	11	12	10	10	11	6	11	7	7	4	5	6	9	0	
Pattern (D)	13	16	7	7	11	10	13	6	5	6	5	4	13	9	0

Note: The numbers shown are the number of times that the congressmen voted differently on 19 environmental bills. R = Republican party, D = Democratic party.

Table 11.6 Coordinates of 15 congressmen obtained
from a three dimensional nonmetric multidimensional
scaling based on voting behavior

Congressman	Dimension		
	1	2	3
Hunt (R)	0.33	0.00	0.09
Sandman (R)	0.26	0.26	0.18
Howard (D)	−0.21	0.05	0.11
Thompson (D)	−0.12	0.22	−0.03
Frelinghuysen (R)	0.20	−0.06	−0.24
Forsythe (R)	0.13	−0.13	−0.06
Widnall (R)	0.33	0.00	−0.11
Roe (D)	−0.21	−0.05	0.09
Helstoski (D)	−0.22	0.02	−0.01
Rodino (D)	−0.16	−0.07	0.00
Minish (D)	−0.16	−0.03	−0.02
Rinaldo (R)	−0.18	0.01	−0.01
Maraziti (R)	0.19	−0.20	0.10
Daniels (D)	−0.02	−0.09	0.03
Pattern (D)	−0.16	0.05	−0.12

Note: R = Republican party, D = Democratic party.

The second analysis was carried out using a classical nonmetric scaling, so that the regression of d_{ij} on δ_{ij} was assumed to be monotonic only. In this case, the stress values for two-, three-, and four-dimensional solutions were found to be 0.113, 0.066, and 0.044, respectively. The distinctly lower stress values for nonmetric scaling suggest that this is preferable to metric scaling for these data, and the three-dimensional nonmetric solution has only slightly more stress than the four-dimensional solution. This three-dimensional nonmetric solution is, therefore, the one that will be considered in more detail. Table 11.6 shows the coordinates of the congressmen for the three-dimensional solution, and plots of the congressmen against the three dimensions are shown in Figure 11.5.

From Figure 11.5, it is clear that Dimension 1 is largely reflecting party differences, because the Democrats fall on the left-hand side of the figure, and the Republicans, other than Rinaldo, fall on the right-hand side.

To interpret Dimension 2, it is necessary to consider what it is about the voting of Sandman and Thompson, who have the highest two scores, that contrasts with Maraziti and Forsythe, who have the two lowest scores. This points to the number of abstentions from voting. Sandman abstained from nine votes, and Thompson abstained from six votes, while individuals with low scores on Dimension 2 voted all or most of the time.

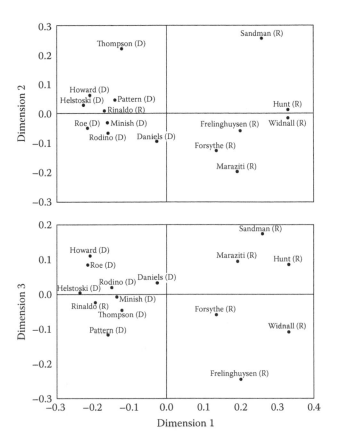

Figure 11.5 Plots of congressmen against the three dimensions obtained from a nonmetric multidimensional scaling.

Dimension 3 appears to have no simple or obvious interpretation, although it must reflect certain aspects of differences in voting patterns. It suffices to say that the analysis has produced a representation of the congressmen in three dimensions that indicates how they relate with regard to voting on environmental issues.

Figure 11.6 shows a plot of the distances between the congressmen for the original data (the disparities) against the points on the derived configuration. This indicates how well the three-dimensional model fits the data. A perfect representation of the data would show the data distances always increasing with the configuration distances. This is not obtained. Instead, there is a range of configuration distances associated with each of the discrete data distances. For example, data distances of 5 correspond to configuration distances from about 0.10 to about 0.16.

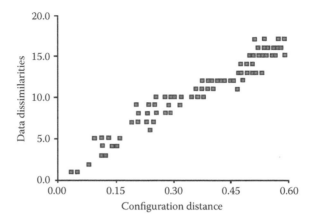

Figure 11.6 The original data distances between the congressmen plotted against the distances obtained for the fitted configuration.

11.3   Computer programs

The Appendix to this chapter provides details on R packages that can be used for multidimensional scaling analyses. Some of the standard statistical packages also include a multidimensional scaling option, but in general, it can be expected that different packages may use slightly different algorithms, and therefore may not give exactly the same results. However, with good data, it can be hoped that the differences will not be substantial.

11.4   Further reading

The classic book by Kruskal and Wish (1978) provides a short introduction to multidimensional scaling. More comprehensive treatments of the theory and applications of this topic and related topics are provided by Cox and Cox (2000) and Borg and Groenen (2005).

Exercise

Consider the data on percentages employed in 26 countries in Europe in Table 1.5. From these data, construct a matrix of Euclidean distances between the countries using Equation 5.1. Carry out nonmetric multidimensional scaling using this matrix to find out how many dimensions are needed to represent the countries in a manner that reflects differences between their employment patterns.

References

Borg, I. and Groenen, P. (2005). *Modern Multidimensional Scaling: Theory and Applications*. 2nd Edn. New York: Springer.

Cox, T.F. and Cox, M.A.A. (2000). *Multidimensional Scaling*. 2nd Edn. Boca Raton, FL: Chapman and Hall/CRC.

Hintze, J. (2012). *NCSS 8. NCSS LLC*. (www.ncss.com.)

Kruskal, J.B. (1964a). Multidimensional scaling by optimizing goodness of fit to a nonmetric hypothesis. *Psychometrics* 29: 1–27.

Kruskal, J.B. (1964b). Nonmetric multidimensional scaling: A numerical method. *Psychometrics* 29: 115–29.

Kruskal, J.B. and Wish, M. (1978). *Multidimensional Scaling*. Thousand Oaks, CA: Sage.

Romesburg, H.C. (2004). *Cluster Analysis for Researchers*. Morrisville, NC: Lulu.com.

Togerson, W.S. (1952). Multidimensional scaling. 1. Theory and method. *Psychometrics* 17: 401–19.

Appendix: Multidimensional scaling in R

When choosing the proper R command for multidimensional scaling, it is important to recall the difference between the classical metric and the nonmetric versions of this method. The classical multidimensional scaling is also known as *principal coordinates analysis*, a topic covered in Chapter 12. The standard acronym for this analysis is *principal coordinate ordination* (PCO). For this analysis, cmdscale() is the main R function that has to be run, as described in the Appendix for Chapter 12. The present chapter is concerned with Kruskal's nonmetric multidimensional scaling (NMDS). This is implemented as the R function isoMDS() in the package MASS (Venables and Ripley, 2002). The main argument of this function is a distance matrix that is a full, symmetric matrix or is created by the function dist(). As an example, assuming that sym.mat is a symmetric matrix, the command

$$NMDS.obj < -isoMDS(sym.mat)$$

will store the results of the NMDS in the object NMDS.obj. The defaults assumed when running isoMDS with a distance matrix as the only argument are k (the number of dimensions, which is 2), maxit (the number of evaluations of stress [Equation 11.1] until convergence is 50), and tol (the tolerance is 1×10^{-3}; that is, once two consecutive values of the stress in the iterative process differ by 1×10^{-3} or less, the process stops, and an outcome for the analysis is produced). For a particular dimension chosen, sometimes the defaults for the number of iterations and the tolerance are somewhat slack, and a better multidimensional scaling configuration can be achieved if the user increases the number of iterations and/or decreases the tolerance. This is illustrated in the R script, accessible from the book's website, that has been written to carry out the analysis for the data in Example 11.1.

According to Step 1 in the procedure for NMDS given in Section 11.2, isoMDS() uses an initial configuration of objects, either supplied by the user or generated by a PCO of the data, which is the default, with an automated intervention of the cmdscale() function. The object NMDS. obj produced by isoMDS() is a two-object list. The first is a scalar for the stress, namely NMDS.obj$stress, given as percent. The second is the matrix NMDS.obj$points, which carries the coordinates of the sampling units in the reduced dimension chosen. Any pair of dimensions can be selected and the corresponding coordinates displayed in a two-dimensional plot like Figures 11.4 and 11.5 using the plot() command. Also, the original distance matrix, sym.mat, and the matrix NMDS.obj$points can be used as arguments of the function called Shepard() to produce a scatterplot similar to that in Figure 11.6. See the R scripts for this chapter for details about the way to obtain this plot, using Example 11.1 as an illustration.

One alternative to isoMDS is offered by metaMDS, which is part of the vegan library (Oksanen et al., 2016). The developers of the metaMDS routine emphasize that it allows greater automation of the multidimensional scaling process than isoMDS. Also, metaMDS tries to eliminate the inaccuracies of isoMDS when this is run with the defaults and convergence is not guaranteed. Actually, metaMDS uses isoMDS in its calculations, but metaMDS is more versatile because it allows random starts of the function initMDS for the object configuration (Step 1 in Section 11.2), and scaling and rotation of the results (function postMDS) are available. This makes the routine more similar to the eigenvalue methods, as the final configuration of the NMDS is followed by a rotation via principal components analysis, so that, for example, the NMDS axis 1 reflects the principal source of variation.

A third option in multidimensional scaling analysis uses the application of optimization (majorization) algorithms, where a particular objective function such as the stress has to be minimized. A collection of R commands following the principle of majorization has been put together in the smacof package for Scaling by MAjorizing a COmplicated Function (de Leeuw and Mair, 2009). In particular, with smacofSym() or the wrapper function mds(), it is possible to perform multidimensional scaling on any symmetric dissimilarity matrix. When the option type="ordinal" is written as an argument of this function, a nonmetric multidimensional scaling algorithm is used. For details, see the paper by de Leeuw and Mair (2009) or the R documentation for the smacof package. Finally, the reader may like to run the R scripts that are accessible at the book's website containing the commands metaMDS() and mds() as alternative ways to produce the NMDS analyses illustrated in Examples 11.1 and 11.2.

References

de Leeuw, J. and Mair, P. (2009). Multidimensional scaling using majorization: SMACOF in R. *Journal of Statistical Software* 31: 1–30.

Oksanen, J., Blanchet, F.G., Friendly, M., Kindt, R., Legendre, P., McGlinn, D., Minchin, P.R., et al. (2016). vegan: Community Ecology Package. R package version 2.4-0. http://CRAN.R-project.org/package=vegan

Venables, W.N. and Ripley, B.D. (2002). *Modern Applied Statistics.* 4th Edn. New York: Springer.

chapter twelve

Ordination

12.1   The ordination problem

The word *ordination* for a biologist means essentially the same as *scaling* does for a social scientist. Both words describe the process of producing a small number of variables that can be used to describe the relationship between a group of objects, starting either from a matrix of distances or similarities between the objects, or from the values of some variables measured on each object. From this point of view, many of the methods that have been described in earlier chapters can be used for ordination, and some of the examples have been concerned with this process. In particular, plotting female sparrows against the first two principal components of size measurements (Example 5.1), plotting European countries against the first two principal components for employment variables (Example 5.2), producing a map of the South Island of New Zealand from a table of distances between towns by multidimensional scaling (Example 11.1), and plotting New Jersey congressmen against axes obtained by multidimensional scaling based on voting behavior (Example 11.2) are all examples of ordination. In addition, discriminant function analysis can be thought of as a type of ordination that is designed to emphasize the differences between objects in different groups, while canonical correlation analysis can be thought of as a type of ordination that is designed to emphasize the relationships between two groups of variables measured on the same objects.

Although ordination can be considered to cover a diverse range of situations, in biology it is most often used as a means of summarizing the relationships between different species as determined from their abundances at a number of different locations or, alternatively, as a means of summarizing the relationships between different locations on the basis of the abundances of different species at those locations. It is this type of application that is considered particularly in the present chapter, although the examples involve archaeology as well as biology. The purpose of the chapter is to give more examples of the use of principal components analysis and multidimensional scaling in this context, and to describe the methods of principal coordinates analysis and correspondence analysis that have not been covered in earlier chapters.

12.2 Principal components analysis

Principal components analysis has already been discussed in Chapter 6. It may be recalled that it is a method whereby the values for variables X_1, X_2, ..., X_p measured on each of n objects are used to construct principal components Z_1, Z_2, ..., Z_p that are linear combinations of the X variables and are such that Z_1 has the maximum possible variance; Z_2 has the largest possible variance, conditional on it being uncorrelated with Z_1; Z_3 has the maximum possible variance, conditional on it being uncorrelated with both Z_1 and Z_2; and so on. The idea is that it may be possible for some purposes to replace the X variables with a smaller number of principal components, with little loss of information.

In terms of ordination, it can be hoped that the first two principal components are sufficient to describe the differences between the objects, because then a plot of Z_2 against Z_1 provides what is required. It is less satisfactory to find that three principal components are important, but a plot of Z_2 against Z_1 with values of Z_3 indicated may be acceptable. If four or more principal components are important, then, of course, a good ordination is not obtained, at least as far as a graphical representation is concerned.

Example 12.1: Plant species in the Steneryd Nature Reserve

Table 9.7 shows the abundances of 25 plant species on 17 plots from a grazed meadow in Steneryd Nature Reserve in Sweden, as described in Exercise 1 of Chapter 9, which was concerned with using the data for cluster analyses. Now, it is an ordination of the plots that will be considered, so that the variables for principal components analysis are the abundances of the plant species. In other words, in Table 9.7 the objects of interest are the plots (columns), and the variables are the species (rows).

Because there are more species than plots, the number of nonzero eigenvalues in the correlation matrix is determined by the number of plots. In fact, there are 16 nonzero eigenvalues, as shown in Table 12.1. The first three components account for about 69% of the variation in the data, which is not a particularly high amount. The coefficients for the first three principal components are shown in Table 12.2. They are all contrasts between the abundance of different species that may well be meaningful to a botanist, but no interpretations will be attempted here.

Figure 12.1 shows a draftsman's diagram of the plot number (1–17) and the first three principal components. It is noticeable that the first component is closely related to the plot number. This reflects the fact that the plots are in order of the abundance in the plots of species with a high response to light and a low response to moisture, soil reaction, and nitrogen. Hence, the analysis has at least been able to detect this trend.

Table 12.1 Eigenvalues from a principal components analysis of the data in Table 9.7 treating the plots as the objects of interest and the species counts as the variables

Component	Eigenvalue	% of Total	Cumulative %
1	8.79	35.17	35.17
2	5.59	22.34	57.51
3	2.96	11.82	69.33
4	1.93	7.72	77.04
5	1.58	6.32	83.37
6	1.13	4.52	87.89
7	0.99	3.97	91.86
8	0.55	2.18	94.04
9	0.40	1.60	95.64
10	0.35	1.40	97.04
11	0.20	0.78	97.82
12	0.18	0.70	98.53
13	0.13	0.51	99.04
14	0.12	0.46	99.50
15	0.07	0.30	99.80
16	0.05	0.20	100.00
Total	25.00	100.00	

Table 12.2 The first three principal components for the data in Table 9.7

Species	Z_1	Z_2	Z_3
Festuca ovina	0.30	0.01	−0.07
Anemone nemorosa	−0.25	0.02	−0.19
Stallaria holostea	−0.20	0.20	−0.19
Agrostis tenuis	0.17	0.14	0.01
Ranunculus ficaria	−0.11	−0.32	−0.07
Mercurialis perennis	−0.08	−0.31	0.02
Poa pratenis	−0.11	0.32	−0.11
Rumex acetosa	−0.01	0.34	0.23
Veronica chamaedrys	−0.15	0.36	−0.06
Dactylis glomerata	−0.23	0.15	0.18
Fraxinus excelsior (juv.)	−0.26	−0.11	0.17
Saxifraga granulate	0.13	0.24	0.23
Deschampsia flexuosa	−0.05	0.12	−0.45
Luzula campestris	0.28	0.09	0.00
Plantago lanceolata	0.27	0.11	0.26

(Continued)

Table 12.2 (Continued) The first three principal components for
the data in Table 9.7

Species	Z_1	Z_2	Z_3
Festuca rubra	−0.03	0.23	0.19
Hieracium pilosella	0.27	−0.02	0.05
Geum urbanum	−0.20	−0.18	0.29
Lathyrus montanus	−0.15	0.26	−0.19
Campanula persicifolia	−0.21	0.18	0.07
Viola riviniana	−0.24	0.17	0.11
Hepatica nobilis	−0.21	0.03	0.34
Achillea millefolium	0.29	0.03	0.10
Allium sp.	−0.18	−0.12	0.36
Trifolim repens	0.21	0.11	0.22

Note: The values shown are for the coefficients of the standardized species
abundances, with means of zero and standard deviations of one.

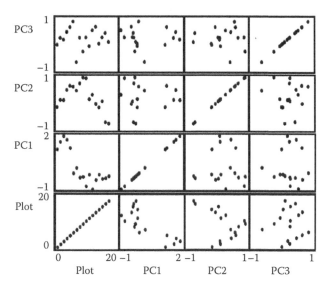

Figure 12.1 Draftsman's diagram for the ordination of 17 plots from Steneryd
Nature Reserve.

Example 12.2: Burials in Bannadi

For a second example of principal components ordination, the
data shown in Table 9.8 concerning grave goods from a cemetery
in Bannadi, northeast Thailand, will be considered. The table, supplied by Professor C.F.W. Higham, shows the presence or absence
of 38 different types of article in each of 47 burials, with additional

information on whether the body was of an adult male, an adult female, or a child. In Exercise 2 of Chapter 9, it was suggested that cluster analysis should be used to study the relationships between the burials. Now, ordination is considered with the same end in mind. For a principal components analysis, the burials are the objects of interest, and the 38 types of grave goods provide the 0–1 variables to be analyzed. These variables were standardized before use so that the analysis was based on their correlation matrix.

In a situation like this, where only presence and absence data are available, it is common to find that a fairly large number of principal components are needed to account for most of the variation in the data. This is certainly the case here, with 11 components needed to account for 80% of the variance and 15 required to account for 90% of the variance. Obviously, there are far too many important principal components for a satisfactory ordination.

For this example, only the first four principal components will be considered, with the understanding that much of the variation in the original data is not accounted for. In fact, the four components correspond to eigenvalues of 5.29, 4.43, 3.65, and 3.34, while the total of all the eigenvalues is 38 (the number of types of articles). Thus, these components account for 13.9%, 11.6%, 9.6%, and 8.8%, respectively, of the total variance, and between them they account for 43.9% of the variance.

The coefficients of the standardized presence-absence variables are shown in Table 12.3 with the largest values (arbitrarily set at an absolute value greater than 0.2) underlined. To aid in interpretation, the signs of the coefficients have been reversed if necessary from what was given by the computer output to ensure that the values of all the components are positive for burial B48, which has the largest number of items present. This is allowable, because switching the signs of all the coefficients for a component does not change the percentage of variation explained by the component, and the direction of the signs is merely an accidental outcome of the numerical methods used to find the eigenvectors of the correlation matrix.

From the large coefficients of Component 1, it can be seen that this is indicating the presence of articles type 9, 10, 16, 18, 19, 20, 23, 25, 26, 30, 32, 34, and 37, and the absence of articles type 3, 5, 6, 14, 28, and 29. There is no grave with exactly this composition, but the component measures the extent to which each of the graves matches this model. The other components can also be interpreted in a similar way from the coefficients in Table 12.3.

Figure 12.2 shows a draftsman's plot of the total number of goods, the type of body, and the first four principal components. From studying this, it is possible to draw some conclusions about the nature of the graves. For example, it seems that male graves tend to have low values and female graves to have high values for principal component 1, possibly reflecting a difference in grave goods associated with sex. Also, grave B47 has an unusual composition in comparison with the other graves. However, the fact that four principal components are being considered makes a simple interpretation of the results difficult.

Table 12.3 Coefficients of standardized presence–absence data
for the first four principal components of the Bannadi data

Article	PC1	PC2	PC3	PC4
1	0.01	−0.02	0.01	−0.00
2	−0.09	−0.04	−0.02	**0.52**
3	**−0.23**	**0.39**	−0.01	−0.03
4	−0.09	−0.04	−0.02	**0.52**
5	**−0.23**	**0.39**	−0.01	−0.03
6	**−0.23**	**0.39**	−0.01	−0.03
7	−0.02	−0.05	−0.02	−0.02
8	−0.03	−0.02	−0.02	−0.02
9	0.17	0.12	0.33	0.06
10	0.15	0.09	0.04	0.07
11	0.05	0.03	0.03	−0.02
12	0.12	0.04	0.11	0.05
13	−0.01	−0.05	−0.02	−0.09
14	**−0.23**	**0.39**	−0.01	−0.03
15	0.00	−0.00	0.03	−0.01
16	0.17	0.12	**0.33**	0.06
17	−0.01	−0.05	−0.02	−0.09
18	**0.22**	0.15	**−0.38**	0.04
19	**0.22**	0.15	**−0.38**	0.04
20	**0.22**	0.15	**−0.38**	0.04
21	−0.09	−0.04	−0.02	**0.52**
22	−0.00	−0.04	0.01	0.01
23	**0.27**	0.17	**−0.24**	0.08
24	0.03	0.03	0.05	−0.07
25	**0.26**	0.15	**0.28**	0.11
26	**0.26**	0.15	**0.28**	0.11
27	0.08	0.02	0.02	0.04
28	**−0.22**	0.19	−0.02	0.26
29	−0.17	0.17	0.01	−0.08
30	0.17	0.11	0.00	−0.05
31	0.08	0.03	0.18	−0.04
32	**0.27**	0.14	0.04	0.03
33	−0.02	−0.00	0.06	−0.12
34	**0.23**	0.09	−0.10	0.03
35	0.04	0.01	0.17	0.14
36	−0.07	0.15	0.17	−0.08
37	**0.26**	0.11	0.07	0.02
38	0.12	**0.22**	0.05	−0.05

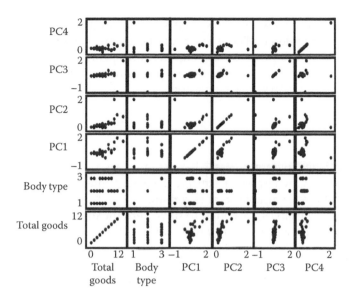

Figure 12.2 Draftsman's diagram for 47 Bannadi graves. The variables plotted are the total number of different types of goods, the type of remains (1 = adult male, 2 = adult female, 3 = child) and the first four principal components.

12.3   Principal coordinates analysis

Principal coordinates analysis is similar to metric multidimensional scaling as discussed in Chapter 11. Both methods start with a matrix of similarities or distances between a number of objects and endeavor to find ordination axes. However, they differ in the numerical approach that is used. Principal coordinates analysis uses an eigenvalue approach that can be thought of as a generalization of principal components analysis. However, multidimensional scaling, at least as defined in this book, attempts instead to minimize the stress, where this is a measure of the extent to which the positions of objects in a t-dimensional configuration fail to match the original distances or similarities after appropriate scaling.

To see the connection between principal coordinates analysis and principal components analysis, it is necessary to recall some of the theoretical results concerning principal components analysis from Chapter 6, and to use some further results that are mentioned here for the first time. In particular:

1. The ith principal component is a linear combination

$$Z_i = a_{i1}X_1 + a_{i2}X_2 + \ldots + a_{ip}X_p$$

of the variables X_1, X_2, ..., X_p that are measured on each of the objects being considered. There are p of these components, and the coefficients a_{ij} are given by the eigenvector $\mathbf{a}_i$ corresponding to the ith largest eigenvalue λ_i of the sample covariance matrix $\mathbf{C}$ of the X variables. That is to say, the equation

$$\mathbf{Ca}_i = \lambda_i \mathbf{a}_i \qquad (12.1)$$

is satisfied where $\mathbf{a}_i' = (a_{i1}, a_{i2}, ..., a_{ip})$. Also, the variance of Z_i is $\mathrm{Var}(Z_i) = \lambda_i$, where this is zero if Z_i corresponds to a linear combination of the X variables that is constant.

2. If the X variables are coded to have zero means in the original data, then the p by p covariance matrix $\mathbf{C}$ has the form

$$\mathbf{C} = \begin{bmatrix} \sum x_{i1}^2 & \sum x_{i1}x_{i2} & \cdots & \sum x_{i1}x_{ip} \\ \sum x_{i2}x_{i1} & \sum x_{i2}^2 & \cdots & \sum x_{i2}x_{ip} \\ \cdot & \cdot & & \cdot \\ \cdot & \cdot & & \cdot \\ \sum x_{ip}x_{i1} & \sum x_{ip}x_{i2} & \cdots & \sum x_{ip}^2 \end{bmatrix} / (n-1)$$

where there are n objects, x_{ij} is the value of X_j for the ith object, and the summations are for i from 1 to n. Hence,

$$\mathbf{C} = \mathbf{X'X}/(n-1) \qquad (12.2)$$

where

$$\mathbf{X} = \begin{bmatrix} x_{i1} & x_{i2} & \cdots & x_{1p} \\ x_{21} & x_{22} & \cdots & x_{2p} \\ \cdot\cdot & & & \cdot \\ \cdot\cdot & & & \cdot \\ x_{n1} & x_{n2} & \cdots & x_{np} \end{bmatrix}$$

is a matrix containing the original data values.

3. The symmetric n by n matrix

$$S = XX' = \begin{bmatrix} \sum x_{1j}^2 & \sum x_{1j}x_{2j} & \cdots & \sum x_{1j}x_{nj} \\ \sum x_{2j}x_{1j} & \sum x_{2j}^2 & \cdots & \sum x_{2j}x_{nj} \\ \cdot & \cdot & & \cdot \\ \cdot & \cdot & & \cdot \\ \sum x_{nj}x_{1j} & \sum x_{nj}x_{2j} & \cdots & \sum x_{nj}^2 \end{bmatrix}$$

where the summations are for j from 1 to p, can be thought of as containing measures of the similarities between the n objects being considered. This is not immediately apparent, but is justified by considering the squared Euclidean distance from object i to object k, which is

$$d_{ik}^2 = \sum_{j=1}^{p} \left(x_{ij} - x_{kj} \right)^2$$

Expanding the right-hand side of this equation shows that

$$d_{ik}^2 = s_{ii} + s_{kk} - 2s_{ik} \tag{12.4}$$

where s_{ik} is the element in the ith row and kth column of XX'. It follows that s_{ik} is a measure of the similarity between objects i and k, because increasing s_{ik} means that the distance d_{ik} between the objects is decreased. Further, it is seen that s_{ik} takes the maximum value of $(s_{ii} + s_{kk})/2$ when $d_{ik} = 0$, which occurs when the objects i and k have identical values for the variables X_1 to X_p.

4. If the matrix

$$Z = \begin{bmatrix} z_{i1} & z_{i2} & \cdots & z_{1p} \\ z_{21} & z_{22} & \cdots & z_{2p} \\ \cdot & \cdot & & \cdot \\ \cdot & \cdot & & \cdot \\ z_{n1} & z_{n2} & \cdots & z_{np} \end{bmatrix}$$

contains the values of the p principal components for the n objects being considered, then this can be written in terms of the data matrix X as

$$Z = XA' \tag{12.5}$$

where the ith row of $\mathbf{A}$ is $\mathbf{a}_i'$, the ith eigenvector of the sample covariance matrix $\mathbf{C}$. It is a property of $\mathbf{A}$ that $\mathbf{A}'\mathbf{A}=\mathbf{I}$, that is, the transpose of $\mathbf{A}$ is the inverse of $\mathbf{A}$. Thus, postmultiplying both sides of Equation 12.5 by $\mathbf{A}$ gives

$$\mathbf{X} = \mathbf{ZA} \tag{12.6}$$

This statement of results has been lengthy, but it has been necessary to explain principal coordinates analysis in relationship to principal components analysis. To see this relationship, note that from Equations 12.1 and 12.2

$$\mathbf{X}'\mathbf{Xa}_i/(n-1) = \lambda_i\mathbf{a}_i$$

Then, premultiplying both sides of this equation by $\mathbf{X}$ and using Equation 12.3 gives

$$\mathbf{S}(\mathbf{Xa}_i) = (n-1)\lambda_i(\mathbf{Xa}_i)$$

or

$$\mathbf{Sz}_i = (n-1)\lambda_i\mathbf{z}_i \tag{12.7}$$

where $\mathbf{z}_i = \mathbf{Xa}_i$ is a vector of length n, which contains the values of Z_i for the n objects being considered. Therefore, the ith largest eigenvalue of the similarity matrix $\mathbf{S}=\mathbf{X}'\mathbf{X}$ is $(n-1)\lambda_i$, and the corresponding eigenvector gives the values of the ith principal component for the n objects.

Principal coordinates analysis consists of applying Equation 12.7 to an n by n matrix $\mathbf{S}$ of similarities between n objects that is calculated using any of the many available similarity indices. In this way, it is possible to find the principal components corresponding to $\mathbf{S}$ without necessarily measuring any variables on the objects of interest. The components will have the properties of principal components, and, in particular, will be uncorrelated for the n objects.

Applying principal coordinates analysis to the matrix $\mathbf{XX}'$ will give essentially the same ordination as a principal components analysis on the data in $\mathbf{X}$. The only difference will be in terms of the scaling given to the components. In principal components analysis, it is usual to scale the ith component to have the variance λ_i, but with a principal coordinates analysis, the component would usually be scaled to have a variance of $(n-1)\lambda_i$. This difference is immaterial, because it is only the relative values of objects on ordination axes that are important.

There are two complications that can arise in a principal coordinates analysis that must be mentioned. They occur when the similarity matrix analyzed does not have all the properties of a matrix calculated from data using the equation $S = XX'$.

First, from Equation 12.3, it can be seen that the sums of the rows and columns of XX' are all zero. For example, the sum of the first row is

$$\sum x_{1j}^2 + \sum x_{1j} x_{2j} + \ldots + \sum x_{1j} x_{nj} = \sum x_{ij} \left(x_{1j} + x_{2j} + \ldots + x_{nj} \right)$$

where the summations are for j from 1 to p. This is zero, because $x_{1j} + x_{2j} + \ldots + x_{nj}$ is n times the mean of X_j, and all the X variables are assumed to have zero means. Hence, it is required that the similarity matrix S should have zero sums for rows and columns. If this is not the case, then the initial matrix can be double-centered by replacing the element s_{ik} in row i and column k by $s_{ik} - s_i - s_{.k} + s$ where s_i is the mean of the ith row of S, $s_{.k}$ is the mean of the kth column of S, and s is the mean of all the elements in S. The double-centered similarity matrix will have zero row and column means and is therefore more suitable for the analysis.

The second complication is that some of the eigenvalues of the similarity matrix may be negative. This is disturbing, because the corresponding principal components appear to have negative variances! However, the truth is just that the similarity matrix could not have been obtained by calculating $S = XX'$ for any data matrix. With ordination, only the components associated with the largest eigenvalues are usually used, so that a few small negative eigenvalues can be regarded as being unimportant. Large negative eigenvalues suggest that the similarity matrix being used is not suitable for ordination.

Computer programs for principal coordinates analysis sometimes offer the option of starting with either a distance matrix or a similarity matrix. If a distance matrix is used, then it can be converted to a similarity matrix by transforming the distance d_{ik} to the similarity measure $s_{ik} = -d_{ik}^2/2$, as suggested by Equation 12.4.

Example 12.3: Plant species in the Steneryd Nature Reserve (revisited)

As an example of the use of principal coordinates analysis, the data considered in Example 12.1 on species abundances on plots in Steneryd Nature Reserve were reanalyzed using Manhattan distances between plots. That is, the distance between plots i and k was measured by $d_{ik} = \sum |x_{ij} - x_{kj}|$, where the summation is for j over the 25 species and x_{ij} denotes the abundance of species j on plot i, as given in Table 9.7. Similarities were calculated as $s_{ik} = -d_{ik}^2/2$ and then double-centered before eigenvalues and eigenvectors were calculated.

The first two eigenvalues of the similarity matrix were found to be 97,638.6 and 55,659.5, which account for 47.3% and 27.0% of the sum of the eigenvalues, respectively. On the face of it, the first two components, therefore, give a good ordination, with 74.3% of the variation accounted for. The third eigenvalue is much smaller at 12,488.2 and only accounts for 6.1% of the total.

Figure 12.3 shows a draftsman's diagram of the plot number and the first two components. Both components show a relationship with the plot number, which, as noted in Example 12.1, is itself related to the response of the different species to environmental variables. Actually, a comparison of this draftsman's diagram with the six graphs in the bottom left-hand corner of Figure 12.1 shows that the first two axes from the principal coordinates analysis are really very similar to the first two principal components apart from a difference in scaling.

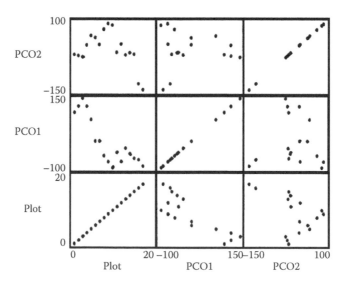

Figure 12.3 Draftsman's diagram for the ordination of 17 plots from Steneryd Nature Reserve based on a principal coordinates analysis on Manhattan distances between plots. The three variables are the plot number and the first two components (PCO1 and PCO2).

Example 12.4: Burials in Bannadi (revisited)

As an example of a principal coordinates analysis on presence and absence data, consider again the data in Table 9.8 on grave goods in the Bannadi cemetery in northeast Thailand. The analysis started with the matrix of unstandardized Euclidean distances between the 47 burials, so that the distance from grave i to grave k was taken to be $d_{ik} = \sqrt{\{\sum(x_{ij} - x_{kj})^2\}}$, where the summation is for j from 1 to 38, and x_{ij} is 1 if the jth type of article is present in the ith burial, or is otherwise 0. A similarity matrix was then obtained as described in Example 12.3 and double-centered before eigenvalues and eigenvectors were obtained.

The principal coordinates analysis carried out in this manner gives the same result as a principal components analysis using unstandardized values for the X variables (i.e. carrying out a principal components analysis using the sample covariance matrix instead of the sample correlation matrix). The only difference in the results is in the scalings that are usually given to the ordination variables by principal components analysis and principal coordinates analysis.

The first four eigenvalues of the similarity matrix were 24.9, 19.3, 10.0, and 8.8, corresponding to 21.5%, 16.6%, 8.7%, and 7.6%, respectively, of the sum of all the eigenvalues. These components account for a mere 54.5% of the total variation in the data, but this is better than the 43.9% accounted for by the first four principal components obtained from the standardized data (Example 12.2).

Figure 12.4 shows a draftsman's diagram for the total number of goods in the burials, the type of body (adult male, adult female, or child), and the first four components. The signs of the first and fourth components were switched from those shown on the computer output so as to make them have positive values for burial B48, which contained the largest number of different types of grave goods. It can be seen from the diagram that the first component represents total abundance quite closely, but the other components are not related to this variable. Apart from this, the only obvious thing to notice is that one of the burials had a very low value for the fourth component. This is burial B47, which contained eight different types of article, of which four types were not seen in any other burial.

12.4   Multidimensional scaling

Multidimensional scaling has been discussed already in Chapter 11, where this was defined to be an iterative process for finding coordinates for objects on axes, in a specified number of dimensions, such that the distances between the objects match as closely as possible the distances or similarities that are provided in an input data matrix (Section 11.2). The method will not be discussed further in the present chapter except as required to present the results of using it on the two example sets of data that have been considered with the other methods of ordination.

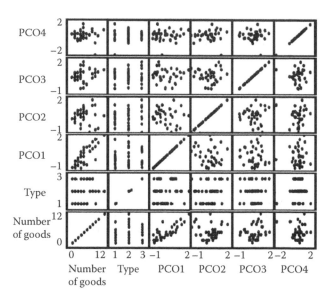

Figure 12.4 Draftsman's diagram for the 47 Bannadi graves. The six variables are the total number of different types of goods in a burial, an indicator of the type of remains (1 = adult male, 2 = adult female, 3 = child), and the first four components from a principal coordinates analysis (PCO1 to PCO4).

Example 12.5: Plant species in the Steneryd Nature Reserve (again)

A multidimensional scaling of the 17 plots for the data in Table 9.7 was carried out using the GenStat program (VSN International Ltd., 2014). This performs a nonmetric type of analysis on a distance matrix, so that the relationship between the data distances and the ordination (configuration) distances is assumed to be only monotonic.

For the example being considered, unstandardized Euclidean distances between the plots were used as input to the program, and a three-dimensional solution was assumed. Figure 12.5 shows the plot numbers plotted against Dimensions 1 and 2, 1 and 3, and 2 and 3. The shows that Dimension 1 is strongly related to the plot number, while Dimension 2 indicates that the central plots differ to some extent from the plots with high and low numbers.

Example 12.6: Burials in Bannadi (again)

The same analysis as used in the last example was also applied to the data on burials at Bannadi shown in Table 9.8. Unstandardized Euclidean distances between the 47 burials were calculated using the 0–1 data in the table as values for 38 variables, and these distances provided the data for GenStat. Figure 12.6 shows the burial numbers plotted against Dimensions 1 and 2, 1 and 3, and 2 and 3. This shows the burials with

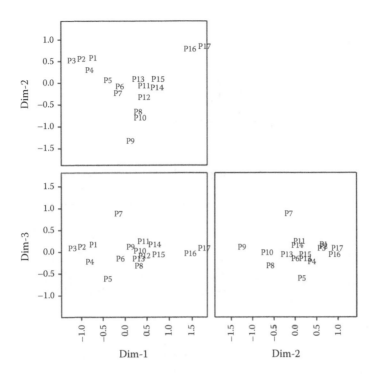

Figure 12.5 Results from the ordination of 17 plots from Steneryd Nature Reserve with nonmetric multidimensional scaling based on Euclidean distances between the plots using unstandardized data, assuming a three-dimensional solution (Dim-1 to Dim-3).

the highest number of goods around the outside of the plots, with the centers of the plots containing the burials with few goods.

12.5   Correspondence analysis

Correspondence analysis as a method of ordination originated in the work of Hirschfeld (1935), Fisher (1940), and a school of French statisticians (Benzecri, 1992). It is a popular method of ordination for plant ecologists and is used in other areas as well.

The method will be explained here in the context of the ordination of p sites on the basis of the abundance of n species, although it can be used equally well on data that can be presented as a two-way table of measures of abundance with the rows corresponding to one type of classification and the columns to a second type of classification.

With sites and species, the situation is as shown in Table 12.4. Here, there are a set of species values $a_1, a_2, \ldots, a_n$ associated with the rows of the table, and a set of site values $b_1, b_2, \ldots, b_p$ associated with the columns of

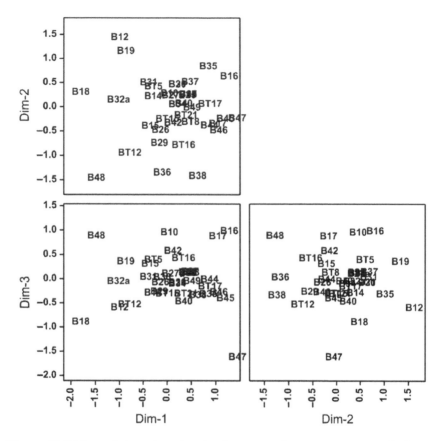

Figure 12.6 Plots of the values for the three axes from non-metric multidimensional scaling using unstandardized Euclidean distances between the Bannadi graves (Dim-1 to Dim-3).

Table 12.4 The abundances (x) of n species at p sites, with the species values (a) and the site values (b)

Species	Site				Row Sum	Species Value
	1	2	...	p		
1	x_{11}	x_{12}	...	x_{1p}	R_1	a_1
2	x_{21}	x_{22}	...	x_{2p}	R_2	a_2
.	.	.	...	.	.	.
.	.	.	...	.	.	.
.	.	.	...	.	.	.
n	x_{n1}	x_{n2}	...	x_{np}	R_n	a_n
Column sum	C_1	C_2	...	C_p		
Site value	b_1	b_2	...	b_p		

the table. One interpretation of correspondence analysis is, then, that it is concerned with choosing species and site values so that they are as highly correlated as possible for the bivariate distribution that is represented by the abundances in the body of the table. That is to say, the site and species values are chosen to maximize their correlation for the distribution in which the number of times that species i occurs at site j is proportional to the observed abundance x_{ij}.

It turns out that the solution to this maximization problem is given by the set of equations

$$a_1 = \{(x_{11}/R_1)b_1 + (x_{12}/R_1)b_2 + \ldots + (x_{1p}/R_1)b_p\}/r$$
$$a_2 = \{(x_{21}/R_2)b_1 + (x_{22}/R_2)b_2 + \ldots + (x_{2p}/R_2)b_p\}/r$$

$$\cdot$$
$$\cdot$$
$$\cdot$$

$$a_n = \{(x_{n1}/R_n)b_1 + (x_{n2}/R_n)b_2 + \ldots + (x_{np}/R_n)b_p\}/r$$

and

$$b_1 = \{(x_{11}/C_1)a_1 + (x_{21}/C_1)a_2 + \ldots + (x_{n1}/C_1)a_n\}/r$$
$$b_2 = \{(x_{12}/C_2)a_1 + (x_{22}/C_2)a_2 + \ldots + (x_{n2}/C_2)a_n\}/r$$

$$\cdot$$
$$\cdot$$
$$\cdot$$

$$b_p = \{(x_{1p}/C_p)a_1 + (x_{2p}/C_p)a_2 + \ldots + (x_{np}/C_p)a_n\}/r$$

where:
   R_i denotes the total abundance of species i
   C_j denotes the total abundance at site j
   r is the maximum correlation being sought

Thus, the ith species value a_i is a weighted average of the site values, with site j having a weight that is proportional to x_{ij}/R_i, and the jth site value b_j is a weighted average of the species values, with species i having a weight that is proportional to x_{ji}/C_j.

The name *reciprocal averaging* is sometimes used to describe the equations just stated, because the species values are (weighted) averages of the site values, and the site values are (weighted) averages of the species values. These equations are themselves often used as the starting point for justifying correspondence analysis as a means of producing species values as a function of site values, and vice versa. It turns out that the equations can be solved iteratively after they have been modified to remove the trivial solution with $a_i = 1$ for all i, $b_j = 1$ for all j, and $r = 1$. However, it is more instructive to write the equations in matrix form to solve them, because this shows that there may be several possible solutions to the equations and that these can be found from an eigenvalue analysis.

In matrix form, the equations just presented become

$$\mathbf{a} = \mathbf{R}^{-1}\mathbf{X}\,\mathbf{b}/r \qquad\qquad (12.8)$$

and

$$\mathbf{b} = \mathbf{C}^{-1}\mathbf{X}'\mathbf{a}/r \qquad\qquad (12.9)$$

where:
  $\mathbf{a}' = (a_1, a_2, \ldots, a_n)$
  $\mathbf{b}' = (b_1\, b_2, \ldots, b_p)$
  $\mathbf{R}$ is an n by n diagonal matrix with R_i in the ith row and ith column
  $\mathbf{C}$ is a p by p diagonal matrix with C_j in the jth row and jth column
  $\mathbf{X}$ is an n by p matrix with x_{ij} in the ith row and jth column

If Equation 12.9 is substituted into Equation 12.8, then, after some matrix algebra, it is found that

$$r^2\left(\mathbf{R}^{\frac{1}{2}}\mathbf{a}\right) = \left(\mathbf{R}^{-\frac{1}{2}}\mathbf{X}\mathbf{C}^{-\frac{1}{2}}\right)\left(\mathbf{R}^{-\frac{1}{2}}\mathbf{X}\mathbf{C}^{-\frac{1}{2}}\right)'\left(\mathbf{R}^{\frac{1}{2}}\mathbf{a}\right) \qquad\qquad (12.10)$$

where:
  $\mathbf{R}^{\frac{1}{2}}$ is a diagonal matrix with $\sqrt{R_i}$ in the ith row and ith column
  $\mathbf{C}^{\frac{1}{2}}$ is a diagonal matrix with $\sqrt{C_j}$ in the jth row and jth column

This shows that the solutions to the problem of maximizing the correlation are given by the eigenvalues of the n by n matrix

$$\left(\mathbf{R}^{-\frac{1}{2}}\mathbf{X}\mathbf{C}^{-\frac{1}{2}}\right)\left(\mathbf{R}^{-\frac{1}{2}}\mathbf{X}\mathbf{C}^{-\frac{1}{2}}\right)'$$

For any eigenvalue λ_k, the correlation between the species and site scores will be $r_k = \sqrt{\lambda_k}$, and the eigenvector for this correlation will be

$$\mathbf{R}^{\frac{1}{2}}\mathbf{a}_k = (\sqrt{R_1}a_{1k}, \sqrt{R_2}a_{2k}, \ldots, \sqrt{R_n}a_{nk})'$$

where a_{ik} are the species values. The corresponding site values can be obtained from Equation 12.9 as

$$\mathbf{b}_k = \mathbf{C}^{-1}\mathbf{X}'\mathbf{a}_k/r_k$$

The largest eigenvalue will always be $r^2 = 1$, giving the trivial solution $a_i = 1$ for all i and $b_j = 1$ for all j. The remaining eigenvalues will be positive or zero and reflect different possible dimensions for representing the relationships between species and sites. These dimensions can be shown to be orthogonal in the sense that the species and site values for one dimension will be uncorrelated with the species and site values in other dimensions for the data distribution of abundances x_{ij}.

Ordination by correspondence analysis involves using the species and site values for the first few largest eigenvalues that are less than 1, because these are the solutions for which the correlations between species values and site values are strongest. It is common to plot both the species and the sites on the same axes, because, as noted earlier in this section, the species values are an average of the site values, and vice versa. In other words, correspondence analysis gives an ordination of both species and sites at the same time.

It is apparent from Equation 12.10 that correspondence analysis cannot be used on data that include a zero row sum, because then the diagonal matrix $R^{-\frac{1}{2}}$ will have an infinite element. By a similar argument, zero column sums are not allowed either. This means that the method cannot be used on the burial data in Table 9.8, because some graves did not contain any articles. However, correspondence analysis can be used with presence and absence data when this problem is not present.

Example 12.7: Plant species in the Steneryd Nature Reserve (yet again)

Correspondence analysis was applied to the data for species abundances in the Steneryd Nature Reserve (Table 9.7). Only the first two dimensions were considered for the ordination plot. Figure 12.7 shows a graph of the species, with abbreviated names, the site numbers, and Dimensions 1 and 2. The ordination of the plots is quite clear, with an almost perfect sequence from Site 17 on the left (S17) to Site 1 on the right, moving around the very distinct arch. The species are interspersed among the sites along the same arch, from *Mercurialis perennis* (Mer-p) on the left, to *Hieracium pilosella* (Hie-p) on the right. A comparison of the figure with Table 9.7 shows that this makes a good deal of sense. For example, *M. perennis* is only abundant on the highest-numbered sites, and *H. pilosella* is only abundant on the lowest-numbered sites.

The arch or horseshoe that appears in the ordination for this example is a common feature in the results of correspondence analysis, which is also sometimes apparent with other methods of ordination. There is sometimes concern that this effect will obscure the nature of the ordination axes, and therefore some attention has been devoted to the development of ways to modify analyses to remove the effect, which is considered to be an artefact of the ordination method.

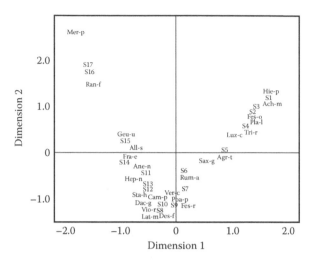

Figure 12.7 Plots of species and site numbers against the first two axes (Dimensions 1 and 2) found by applying correspondence analysis to the data from Steneryd Nature Reserve. The species names have obvious abbreviations and the sites are numbered from S1 to S17.

With correspondence analysis, a method of detrending is usually used, and the resulting ordination method is then called *detrended correspondence analysis* (Hill and Gauch 1980). Adjustments for other methods of ordination exist as well, but seem to receive little use.

12.6   Comparison of ordination methods

Four methods of ordination have been reviewed in this chapter, and it is desirable to be able to state when each should be used. Unfortunately, this cannot be done in an altogether satisfactory way because of the wide variety of different circumstances for which ordination is used. Therefore, all that will be done here is to make some final comments on each of the methods in terms of its utility.

Principal components analysis can only be used when the values for p variables are known for each of the objects being studied. Therefore, this analysis cannot be used when only a distance or similarity matrix is available. When variable values are available and the variables are approximately normally distributed, this method is an obvious choice.

When an ordination is required starting with a matrix of distance or similarities between the objects being studied, it is possible to use either principal coordinates analysis or multidimensional scaling. Multidimensional scaling can be metric or nonmetric, and principal coordinates analysis and metric multidimensional scaling should give similar results. The relative advantages of metric and nonmetric multidimensional scaling will depend very much on the circumstances, but, in general, nonmetric scaling can be expected to give a slightly better fit to the data matrix.

Correspondence analysis was developed for situations where the objects of interest are described by measures of the abundance of different characteristics. When this is the case, this method appears to give ordinations that are relatively easy to interpret. It has certainly found favor with ecologists analyzing data on the abundance of different species at different locations.

12.7   Computer programs

The analyses described in this chapter can be carried out using many standard statistical packages, and there are also some specialized packages developed for specific areas of application such as for analyzing plant distribution data. Also, the Appendix to this chapter explains how the various ordination analyses described in the chapter can be carried out using packages in R.

12.8   Further reading

Suggestions for further reading related to principal components analysis and multidimensional scaling are provided in Chapters 6 and 11, and it is unnecessary to repeat these here. For further discussions and more examples of principal coordinates analysis and correspondence analysis, particularly in the context of plant ecology, see the books by Digby and Kempton (1987), Ludwig and Reynolds (1988), and Jongman et al. (1995). For correspondence analysis, the classic reference is by Greenacre (1984). In addition, there is a short book on correspondence analysis by Clausen (1998) and a very comprehensive book on the same topic by Benzecri (1992).

One important technique not covered in this chapter is canonical ordination, in which the ordination axes are chosen to represent a set of explanatory variables as much as possible. For example, there might be interest in seeing how the distribution of plant species over a number of sites is related to the temperature and soil characteristics at those sites. Discriminant function analysis is one special case of this type of analysis, but a number of other analyses are also possible. See Jongman et al. (1995) for more details.

Exercise

Table 6.6 shows the values for six measurements taken on each of 25 prehistoric goblets excavated in Thailand. The nature of the measurements is shown in Figure 6.3. Use the various methods discussed in this chapter to produce ordinations of the goblets and see which method appears to produce the most useful result.

References

Benzecri, P.J. (1992). *Correspondence Analysis Handbook*. New York: Marcel Dekker.

Clausen, S.E. (1998). *Applied Correspondence Analysis*. Thousand Oaks, CA: Sage.

Digby, P.G.N. and Kempton, R.A. (1987). *Multivariate Analysis of Ecological Communities*. London: Chapman and Hall.

Fisher, R.A. (1940). The precision of discriminant functions. *Annals of Eugenics* 10: 422–9.

Greenacre, M.J. (1984). *Theory and Application of Correspondence Analysis*. London: Academic Press.

Hill, M.O. and Gauch, H.G. (1980). Detrended correspondence analysis, an improved ordination technique. *Vegetatio* 42: 47–58.

Hirschfeld, H.O. (1935). A connection between correlation and contingency. *Proceedings of the Cambridge Philosophical Society* 31: 520–4.

Jongman, R.H.G., ter Braak, C.J.F., and van Tongeren, O.F.F. (1995). *Data Analysis in Community and Landscape Ecology*. Cambridge University Press, Cambridge.

Ludwig, J.A. and Reynolds, J.F. (1988). *Statistical Ecology*. Wiley, New York.

VSN International Ltd. (2014). GenStat, 17th Edn. www.vsni.co.uk

Appendix: Ordination methods in R

Fully commented R scripts, reproducing the results for each example presented in this chapter, are freely available at the book's website. Our strategy in writing all those scripts was to use the simplest R commands for that purpose; at the end of this Appendix, we suggest additional R packages offering a wider range of computational procedures for ordination analysis.

A.1 Principal components and nonmetric multidimensional scaling

It was indicated in this chapter that principal components analysis (Section 12.2) and nonmetric multidimensional scaling (Section 12.4) belong to the collection of eligible methods that a data analyst may apply for producing the ordination of multivariate data. Consequently, we refer the reader to the Appendices in Chapters 6 and 11 to select the R commands needed to generate the reduced set of variables from principal components analysis or nonmetric multidimensional scaling.

A.2 Principal coordinates analysis

Principal coordinates analysis can be run in R with cmdscale, a function whose name stands for classical multidimensional scaling. The main argument for this function is a distance matrix computed by the dist() or any other equivalent function. As an example, given a data frame (say, MV.DATA) containing the variables of interest, and assuming that the Manhattan distance has been chosen to measure the dissimilarities among the sampling units in MV.DATA, the distance matrix can be obtained as

```
Dist.Manh <- dist(MV.DATA, method = "manhattan")
```

Now, suppose you desire the maximum dimension of the reduced space to be $k = 3$, and that eigenvalues should be returned. The cmdscale command that has to be executed is then

```
PCO.object <- cmdscale(Dist.Manh, k = 3, eig = TRUE)
```

The user does not need to worry about transforming the distance matrix into a double-centered similarity matrix as described in Section 12.3. The cmdscale command automatically computes that for you.

In the vegan package, there is the function `capscale()`, which is a constrained version of principal coordinates analysis (PCoA, also known as MDS), but can be used for normal PCoA. It can place both the objects and the variables in a biplot-like figure.

A.3   Correspondence analysis

The R command for correspondence analysis (CA), as described in the present chapter, is `ca()`, from the package ca (Nenadic and Greenacre 2007). This command performs *simple CA*, a term given by the French school to mean that the data of interest in this case are usually a two-way contingency table. The terms *multiple* and *joint correspondence analysis* are used to denote extensions of simple CA to more than two categorical variables. See the paper by Nenadic and Greenacre (2007) for further details.

Once a simple CA is carried out for a data matrix `mat.dat`, for example, by

$$\texttt{object.ca <- ca(mat.dat)}$$

the user can get the complete set of eigenvalues by simply typing

$$\texttt{object.ca}$$

or by means of the command `summary(object.ca)`. In addition, the correlations between row and column scores (or singular values) can be invoked with

$$\texttt{object.ca\$sv}$$

A joint plot of the CA scores (see Figure 12.7) is then produced with the `plot` function

$$\texttt{plot(object.ca).}$$

A.4   Other R packages for ordination

Among the specialized R procedures for ordination analyses, the two particular packages `labdsv` (Roberts, 2016) and vegan (Oksanen et al., 2016) are very good. Both packages offer a diversity of ordination analyses created primarily to suit the computational needs of community ecologists, including principal components analysis, PCoA, and multidimensional scaling. The scope of functions is larger in vegan than in `labdsv`. Thus, vegan allows the selection of two popular correspondence analysis algorithms, single CA and canonical CA, this latter known also as constrained CA, through the single command cca(), while detrended correspondence analysis is invoked with the function decorana(). Several comprehensive books and tutorials about the R functions for ordination analysis using vegan are available. We recommend the book by Borcard et al. (2011) and the tutorial (vignette) given by Oksanen (2016), the creator of the vegan package.

A.5   Draftsman's plots in ordination

A general strategy suggested in the present chapter is to produce draftsman's diagrams as visual aids for drawing conclusions from any ordination analysis. These diagrams show scatter plots for a selected sub-set of variables derived by the multivariate analysis (the first few principal components in principal components analysis, the first few dimensions in PCoA or nonmetric multidimensional scaling, etc.), and they may include a particular set of variables from the original data set, helpful in the interpretation of the ordination results. As seen in Chapter 3, whenever a draftsman's diagram is desired, the R functions pairs() and scatter-plotMatrix() from the car package and splom() from the lattice package are suitable commands that can be added to the R scripts in ordination analysis.

References

Borcard, D., Gillett, F., and Legendre, P. (2011). *Numerical Ecology*. New York: Springer.

Jongman, R.H.G., ter Braak, C.J.F., and van Tongeren, O.F.F. (1995). *Data Analysis in Community and Landscape Ecology*. Cambridge: Cambridge University Press.

Ludwig, J.A. and Reynolds, J.F. (1988). *Statistical Ecology*. New York: Wiley.

Nenadic, O. and Greenacre, M. (2007). Correspondence analysis in R, with two- and three-dimensional graphics: The ca package. *Journal of Statistical Software* 20(3): 1–13.

Oksanen, J. (2016). *Vegan: An Introduction to Ordination*. https://cran.r-project.org/web/packages/vegan/vignettes/intro-vegan.pdf

Oksanen, J., Blanchet, F.G., Friendly, M., Kindt, R., Legendre, P., McGlinn, D., Minchin, P.R., et al. (2016). vegan: Community Ecology Package. R package version 2.4-0. http://CRAN.R-project.org/package=vegan

Roberts, D.W. (2016). labdsv: Ordination and Multivariate Analysis for Ecology. R package version 1.8-0. https://CRAN.R-project.org/package=labdsv

chapter thirteen

Epilogue

13.1 The next step

In writing each edition of this book, the aims have purposely been limited in terms of the content. These aims will have been achieved if someone who has read the previous chapters carefully has a fair idea of what can and what cannot be achieved by the multivariate statistical methods that are most widely used. Our hope is that the book will help many people take the first step in "a journey of a thousand miles."

For those who have taken this first step, the way to proceed further is to gain experience of multivariate methods by analyzing different sets of data and seeing what results are obtained. As with other areas of applied statistics, competence in multivariate analysis requires practice.

Some recent developments in multivariate analysis have been made in the closely related field of data mining, which is concerned with extracting information from very large data sets. This topic has not been considered in this book, but it is an area that should be investigated by anyone dealing with large multivariate data sets. More details will be found in the book by Hand et al. (2001).

13.2 Some general reminders

In developing expertise and familiarity with multivariate analyses, there are a few general points that are worth keeping in mind. Actually, these points are just as relevant to univariate analyses. However, they are still worth emphasizing in the multivariate context.

First, it should be remembered that there are often alternative ways of approaching the analysis of a particular set of data, none of which is necessarily the best. Indeed, several types of analysis may well be carried out to investigate different aspects of the same data. For example, the body measurements of female sparrows given in Table 1.1 can be analyzed by principal components analysis or factor analysis to investigate the dimensions behind body size variation, by discriminant analysis to contrast survivors and nonsurvivors, by cluster analysis or multidimensional scaling to see how the birds group together, and so on.

Second, use common sense. Before embarking on an analysis, consider whether it can possibly answer the questions of interest. Many statistical analyses are carried out because the data are of the right form, irrespective of what light the analyses can throw on a question. At some time or another, most users of statistics find themselves sitting in front of a large pile of computer output with the realization that it tells them nothing that they really want to know.

Third, multivariate analysis does not always work in terms of producing a neat answer. There is an obvious bias in statistical textbooks and articles toward examples in which results are straightforward and conclusions are clear. In real life, this does not happen quite as often. Do not be surprised if multivariate analyses fail to give satisfactory results on the data that you are really interested in! It may well be that the data have a message to give, but the message cannot be read using the somewhat simple models that standard analyses are based on. For example, it may be that variation in a multivariate set of data can be completely described by two or three underlying factors. However, these may not show up in a principal components analysis or a factor analysis because the relationship between the observed variables and the factors is not a simple linear one.

Finally, there is always the possibility that an analysis is dominated by one or two rather extreme observations. These outliers can sometimes be found by simply scanning the data by eye, or by considering frequency tables for the distributions of individual variables. In some cases, a more sophisticated multivariate method may be required. For example, a large Mahalanobis distance from an observation to the mean of all observations is one indication of a multivariate outlier (see Section 5.3), although the truth may just be that the data are not approximately multivariate normally distributed.

It may be difficult to decide what to do about an outlier. If it is due to a recording error or some other definite mistake, then it is fair enough to exclude it from the analysis. However, if the observation is a genuine value, then this is not valid. Appropriate action then depends on the particular circumstances. See Barnett and Lewis (1994) for a detailed discussion of possible approaches to the problem.

What is sometimes effective is to do an analysis with and without the extreme values. If the conclusions are the same, then there is no real problem. It is only if the conclusions depend strongly on the extreme values that they need to be dealt with more carefully.

13.3   Missing values

Missing values can cause more problems with multivariate data than with univariate data. The trouble is that when there are many variables being measured on each individual, it is often the case that one or two of these variables have missing values. It may happen that if individuals with any missing values are excluded from an analysis, then this means excluding quite a large proportion of individuals, which may be completely impractical. For example, in the study of ancient human populations, skeletons are frequently broken and incomplete.

Texts on multivariate analysis are often quite silent on the question of missing values. To some extent, this is because doing something about missing values is by no means a straightforward matter. In practice, computer packages sometimes include a facility for estimating missing values by various methods of varying complexity. One possible approach is, therefore, to estimate missing values and then analyze the data including these estimates, as if they had been complete data in the first place. It seems reasonable to suppose that this procedure will work satisfactorily provided that only a small proportion of values are missing.

For a detailed discussion of methods for dealing with missing data, see the text by Little and Rubin (2002).

References

Barnett, V. and Lewis, T. (1994). *Outliers in Statistical Data*. 3rd Edn. New York: Wiley.

Hand, D., Mannila, H., and Smyth, P. (2001). *Principles of Data Mining*. Cambridge, MA: MIT Press.

Little, R.A. and Rubin, D.B. (2002). *Statistical Analysis with Missing Data*. 2nd Edn. New York: Wiley.

Index